The Future of Tropical Savannas

An Australian Perspective

EDITOR ANDREW ASH

CSIRO Cataloguing-in-Publication

The Future of Tropical Savannas: An Australian Perspective

ISBN 0 643 05784 6

   1. Savannas — Tropics.
   2. Savanna ecology — Tropics.
   I. Ash, Andrew John, 1958–
   II. CSIRO

574.52643

© CSIRO Australia 1996

This book is available from:
   CSIRO Publishing
   P.O. Box 1139
   Collingwood, VIC. 3066
   Australia
   Ph: + (61 3) 9662 7666   Fax: + (61 3) 9662 7555   email: sales@publish.csiro.au

Publisher, Academic & Reference: Kevin Jeans
Cover Design: Melissa Spencer
Text Design: Anita Adams
Production Manager: Jim Quinlan
Cover Photographs: David Curl

The Publisher and Editor would like to thank David Curl for providing and permitting use of his photographs for the front cover and for the section pages.

Preface

R.J. Clements

Chief
CSIRO Division of Tropical Crops and Pastures
306 Carmody Rd,
St Lucia, QLD 4067
Australia

Tropical savannas with their continuous grassy layer and variable overstorey of trees and shrubs are one of the most widespread vegetation types, covering more than 50% of the tropical land mass. They occur on all four tropical continents, in more than 20 countries, between the rainforests and deserts. Tropical savannas cover more than 80% of northern Australia, and industries based there (particularly mining and beef) are major contributors to the Australian economy. However other groups (conservationists, Aboriginal communities, tourism operators and the military) also wish to use the savannas, and future use of the area is subject to intense debate among the users. The most important issues in this debate are how to manage the resources so that options for future use are not closed off, and how to resolve the conflicts which have arisen and will arise among alternative land users.

This book is based on papers presented at a symposium held in Townsville in July 1994. The symposium was organised by the CSIRO Division of Tropical Crops and Pastures, with support and sponsorship from James Cook University of North Queensland, the Office of Northern Development, Queensland Department of Environment and Heritage, CSIRO Division of Soils, CSIRO Division of Wildlife and Ecology, Queensland Department of Lands and Western Australia Department of Conservation and Land Management. The symposium organisers sought to bring together a diverse group of people interested in savannas (users, scientists, and policy makers) and to present them with a range of relevant issues and a workshop framework for discussion. The important points from these discussions are set out in the summary of the workshops in this book.

The symposium was divided into four parts. The introductory section included relevant policy issues and institutional responses to changing resource values, and the role of science in resource management and conflict resolution. In the second section, representatives of the major users of Australian savannas (livestock producers, National Park users, tourism operators, the military, miners and Aboriginal communities) presented their aspirations for future use of the savannas. The third section covered ways to accommodate the different perspectives of the users, and included a number of workshops which allowed participants to discuss in detail ways to develop responsible land use. The final section covered future views of savanna use including changing perceptions of savanna development on the world scene, future industries and management regimes, land use on Cape York Peninsula, and a summative address.

The speakers were mainly Australian and the emphasis was on Australian savannas, but the conflicts in land use have wide relevance to other countries and vegetation types. The views expressed reflect the individuality of the writers as do their writing styles; no attempt has been made to produce needless conformity.

The symposium and this book are an important step in making the future a better one for the savannas and for all savanna users. The organisers deserve great credit for providing us with a perspective that is both novel and progressive, and I am sure you would wish me to acknowledge the members of the Organising Committee. They were John McIvor (Chair), Andrew Ash, Dick Braithwaite, Joel Brown, Ross Coventry, Geoff Cox, Chris Done, Bill Freeland, Dan Gillespie, Bill Gray, Ken Harvey, Raymond Jones, Richard Monypenny, Joe Scanlan, John Taylor and Bill Winter.

Contents

Preface		iii

General Introduction

	Introduction: Tropical Savannas E.J. Lindsay	3
1.	Regional restructuring of the Tropical Savannas: Impacts on Lands, Peoples and Human Settlements J. Holmes	5
2.	Managing Resources and Resolving Conflicts: The Role of Science W. Winter & J. Williams	20
3.	Changing Resource Values in Australia's Tropical Savannas: Priorities in Institutional Reform J. Holmes	28

Savanna Users and their Perspectives

4.	Savanna users and their Perspectives: Grazing Industry J. Stewart	47
5.	National Parks and other Protected Areas P. Bridgewater & D. Walton	54
6.	Tourism in the Tropical Savannas G. Collins	62
7.	The Australian Defence Force and the Future of Tropical Savannas A. Barton & J. McDonald	68
8.	Sustainable Mining in Australia's Tropical Savannas G. Ewing	80
9.	Aboriginal People of the Tropical Savannas: Resource Utilization and Conflict Resolution D. Pearce, A. Jackson & R. Braithewaite	88

Accommodating Different Perspectives

10.	Interactions between Land Uses in Australia's Savannas — It's largely in the mind J. Taylor & R. Braihwaite	107
11.	Mediation and Sustainable Development: Conflict Prevention, Conflict Resolution and Public Participation D. Craig	119
12.	Conflict Management: Achieving Practical Outcomes M. O'Donnell & C. Nolan	125

Looking to the Future

13.	Changing Perceptions on Savanna Development *M. Hadley*	133
14.	What lies ahead for the Tropical Savanna? Industries and Management Regimes *B. Gray*	149
15.	A 2020 Vision for Cape York Peninsula: A story of 40 000 years plus 200 *M. Baird*	159
16.	Towards Responsible Land Use — Conclusions and Outcomes *P. Huthwaite*	165
17.	Summative Address: Perceptions, Issues, Trends and Roles *R.J. Clements*	168

Index   175

General Introduction

Introduction

Tropical Savannas

The Hon. E.J. Lindsay, M.P.

Parliamentary Secretary to the Minister for Industry, Science and Technology
Parliament House
Canberra, ACT 2699
Australia

Thank you for inviting me to open the international symposium and workshop on the future for tropical savannas.

As Parliamentary Secretary to the Minister for Industry, Science and Technology, I have had the privilege of addressing scientists from many disciplines, including remote sensing, physics, marine science and plant-gene technology. But I am sure this audience would understand if I acknowledge that it is especially rewarding to be addressing this symposium on tropical savannas, in my home town of Townsville.

Savannas are of immense importance to tropical regions and are of global significance. They are one of the most widespread vegetation types, covering more than half of the world's tropical land mass. They occur between the rainforest and the deserts on all four tropical continents, in more than 20 countries. Like the tropical rainforests, they are a major reservoir of biodiversity.

The government places a high value on international linkages, particularly with the countries of our region. The Australian Centre for International Agricultural Research and the International Science and Technology program of the Department of Industry, Science and Technology can provide assistance to fund international collaborative research.

Australia's tropical savannas are a vital part of northern Australia. They cover more than one-fifth of the continent, stretching across the top of the country from the Indian Ocean to the Pacific. Although they support only a sparse population of about 200 000 people, they provide the natural resources for economic enterprises that contribute AUD$7.5 billion annually to the national economy. Much of this is in export income from the grazing, mining and tourism industries, and accounts for more than one-fourth of Australia's export merchandise. In addition, Aboriginal landowners and National Parks are assuming an increasingly important role in the use and management of savannas.

Appropriate management is essential for all these uses if we are to prevent degradation of the resources that provide the base for these industries. Ecologically sustainable development is the primary challenge currently facing tropical savanna users and it will remain so for the remainder of this decade.

Last year the Australian Science and Technology Council, ASTEC, reported on research and technology in tropical Australia. The report identified the tropical savanna ecosystem as the highest priority for focused research in tropical Australia. In support, ASTEC cited the increased development pressures on tropical savannas, the damage being caused by weeds and pests, and inadequate knowledge of the ecosystem.

The government recognises that ecologically sustainable development of the tropical savannas will require better knowledge and better management practices. The potential does exist to massively increase the productivity of the savannas. But there are concerns that the pressure for further development could damage long-term productivity. There is a vital role for science in providing solutions for sustainable development to managers.

Following ASTEC's report, an independent working group under Professor Jim McLeod prepared a paper for the Prime Minister's Science and Engineering Council. I note that Dr. Peter Bridgewater and Mr. Bill Gray from the working group will address you during the symposium.

The working group identified the following major priorities for research on the tropical savannas:
i) baseline data on climate, soil, water, plant and animal species;
ii) an understanding of key ecological processes;
iii) methods of prediction, detection, interpretation and rehabilitation to tackle environmental disturbances;
iv) ways of reconciling competing/multiple land uses, involving mining, tourism, pastoralism, conservation, and defence.

The working group also identified an opportunity for the Commonwealth to assist in coordinating improved research management within the savannas.

A joint program has now been agreed on between the Australian Nature Conservation Agency (ANCA) and the Land and Water Resources Research and Development Corporation (LWRRDC). ANCA is focusing on environmental research and the Land and Water Resources Research and Development Corporation (LWRRDC) is looking at research for sustainable resource use for primary industries.

Other organisations with an interest in this area have been approached about possible involvement and government agencies in Western Australia, Queensland and the Northern Territory have expressed strong support for this study.

Under the joint work program, an expanded analysis of current research and development programs is being prepared. And, in addition, a short-term study is to be carried out to identify major issues related to the sustainable use and conservation of Australia's tropical savannas. The program will concentrate on gaps in knowledge, priorities for further research and coordination of research management.

In parallel to the ANCA/LWRRDC initiative, a proposal has been put forward for the establishment of a cooperative research centre (CRC) for the Sustainable Development of Tropical Savannas. This would be located principally in Darwin, with nodes in Townsville and Western Australia. Core participants would be government departments of the Northern Territory, Queensland and Western Australia, Northern Territory University, James Cook University and CSIRO. The main goal of the CRC is to help maintain and expand the sustainable economic development of Australia's savannas. It is very gratifying to see the tropical savanna research community working together, across the two states and one territory of northern Australia and across so many institutional boundaries.

This last round of CRC proposals is certainly going to be very competitive. Over 50 applications have been received, including nine proposals that are environmentally oriented. However the CRC proposal is assessed, I am sure that the framework for continual collaboration that has been established will continue, and that the major needs for research into tropical savannas will be undertaken in a coordinated, consultative and multidisciplinary way.

There are, of course, close linkages between the ANCA/LWRRDC study and the CRC proposal. If the CRC is successful in receiving funding, it will play a major part in implementing the ANCA/LWRRDC strategy for research and development in Australia's tropical savannas.

Improved knowledge is, however, only one aspect of good management. The multiple use of savannas leads to interactions among the users — some positive and some negative — giving rise to conflicts which feature prominently in media reports. Must we have these conflicts? Can they be avoided? How can we resolve them, so that all sides are satisfied? Over the next four days this symposium will provide an ideal forum for discussion of these issues, particularly in the workshops on Thursday.

The federal government has a number of strategies which are relevant to the sustainable development of the savannas. A *National Strategy for Rangeland Management* is being prepared and workshops are currently being held throughout Australia to this end.

The government's policies on sustainable development aim to provide for productive use of tropical savannas and other rangelands by Australians today, while conserving our ecosystems for the benefit of future Australians. Although only a small portion of the population lives in savannas, savanna management is important to all Australians. The government recognises that concerns over land degradation, loss of biodiversity, and social changes occurring in the savannas must therefore be addressed.

Using and managing our tropical savannas is an immense challenge. We must learn to protect and conserve the natural resources and biodiversity in the savannas, while maintaining the use of their resources for the greatest benefit to all Australians.

Chapter 1

Regional Restructuring of the Tropical Savannas: Impacts on Lands, Peoples and Human Settlements

John H. Holmes

Department of Geographical Sciences & Planning,
The University of Queensland,
St. Lucia, QLD. 4072
Australia

Abstract

- More so than any other major bioclimatic zone, Australia's tropical savannas are experiencing a major change in resource values. Lack of success in serving the commodity-oriented values of the industrial era now provides an enhanced opportunity in meeting the people-oriented and environmental values of the postindustrial era.

- This revaluation is most pronounced in the northern savanna zone, where returns to grazing, always modest, are either static or declining while other resource values are increasingly important: Aboriginal use; biodiversity; tourism and recreation.

- New directions are emerging in the identification and appraisal of valued resources, requiring a strong focus on resource-rich locales where multiple-use values generate pressures, involving elements of complementarity, conflict and incompatibility.

- New resource values are contributing to regional restructuring, involving economic, social and demographic change.

- Resource revaluations can be expected to generate steady economic and population growth, focusing on major urban centres and prime resource locations. Steady growth in Aboriginal population can be predicted, with less certain outcomes in geographic distribution and in levels of economic and social welfare. Conversely, the growth rate of the non-Aboriginal population is less predictable, but with more certain distributional and welfare outcomes.

- Apart from the dominant urban centres, notably Darwin, human settlements are likely to remain small, widely scattered, specialised and poorly articulated as economic, social and political communities. Strategic regional planning will need to integrate settlement dynamics with land use change, and should aim for a balance between open, multifunctional (if small) urban centres and closed, single-purpose, enclave settlements.

- Global comparisons suggest that Australia's tropical savannas will pursue an increasingly distinctive development path, differentiated from comparable bioclimatic zones by their incorporation within a western, affluent nation, and from other peripheral zones in western nations by its bioclimatic environment. This opens up possibilities which are yet to be fully recognised.

1. INTRODUCTION

The conference organisers have chosen a timely theme: managing resources and resolving conflicts. In their conference preface, they have emphasized the increasing diversity of uses, the scope for conflict and the need to accommodate different perspectives.

In this paper, I seek to go one step further, and argue that we need to undertake a more far-reaching reconceptualisation of the changing structure of resource use in Australia's tropical savannas, particularly in the northern zone, coincident with the monsoonal tallgrass savannas, as delineated below and originally defined by Mott et al. (1985). We need to recognise that this zone is undergoing a radical revaluation of its resources, in a manner unmatched in settled regions and to an extent unequalled in other Australian rangeland regions. This revaluation is already sharply revealed in the changing ownership and use of the northern savanna lands. This is accompanied by a rapidly changing balance of political and economic power among the interest groups, both within and beyond the tropical savannas, with a growing diversity of interests claiming a role in determining resource issues. These interest-groups are differentiated according to culture and ethnicity (Aboriginal, non-Aboriginal), to priority in resource use (traditional Aboriginal, pastoralism, agriculture, mining, recreation, commercial tourism, conservation, preservation and so on), to locale (within or beyond the savanna region) and to scale (locality, region, state-territory or national). Furthermore, each interest-group is by no means homogeneous in values, goals, expectations and demands.

2. CHANGING RESOURCE VALUES IN A POST-INDUSTRIAL ERA

Revaluation of tropical savanna resources is not merely the outcome of past disappointments with northern resource development, nor from diminished future expectations engendered by technological change which now strongly favours intensification of the more productive agrosystems, thereby further disadvantaging production at the extensive margins. Much more influential are the structural changes occurring widely within western societies, loosely described as the transition from an industrial to a postindustrial society.

The most salient post-industrial change has been away from a commodity-oriented society towards one based on informationmation and knowledge instead of labour, materials, energy and land as the prime factors

Table 1: Industrial and Post-Industrial Eras: Resource Values, Policies, Goals and Institutions Impacting Upon the Tropical Savannas

Attribute	Dominant within Industrial Era	Additional within Post-Industrial Era
Economic Orientation	Market-oriented: income generation	Non-market: human welfare; environmental values; sustainability
Socioeconomic goals	Maximize commodity production	Enhance environmental, amenity, welfare and community values
Marketable Outputs from Natural Resources	Minerals, pastoral products	Tourism, recreation, amenity-focused residential
Major Non-Market Outputs from Natural Resources	(Rarely recognized)	Aboriginal traditional uses, recreation, amenity, existence values
Resource Appraisal Criteria	1. Mineral potential 2. Biological potential: forage and cropping outputs.	1. Biological: bio-diversity, Aboriginal subsistence. 2. Landscape values. 3. Aboriginal traditional values.
Distribution of Land Resource Values	Dispersed: broad-acres pastoral use	Often concentrated at prime sites
Property Rights and Land Tenure	Enhanced private property rights linked to marketable resource outputs	Expansion of Aboriginal land rights and of public tenures tied to non-market resource outputs
Public Investment	Physical infrastructure: roads, ports, telecommunications, irrigation projects	Social infrastructure: education, health, housing, welfare
Mechanisms for Regional Transfer Payments	Commodity subsidies and support, fuel subsidies, cross subsidies within service utilities	Direct payments for welfare: special assistance with education, health, housing and related services
Research Priorities	Production-oriented: selective resource inventories; specialized experimental research	Environmental- and people-oriented: inventories, appraisal and monitoring; multi-disciplinary survey research
Sources of Political Power and Influence	Producer groups: pro-developer advocates	Diversified but with prominent roles for Aboriginals and environmentalists
Contribution of Tropical Savannas to National Goals	Very modest: generally disappointing	High prospects for an enhanced role
Outcomes from Public Programs	Very mixed: many failures	Not yet clear

of production (Brotchie et al. 1987). The discounting of land as a productive resource is tied to the reduced economic significance of raw materials and their diminished contribution to contemporary wealth. Together with technological change and trade protectionism, this has led to a marked devaluation of Australia's extensive pastoral lands. Post-industrial societies are also characterised by a growing emphasis on leisure, amenity and quality-of-life values, together with greater attention to environmental and social justice issues, leading to a significant shift in national goals as well as international covenants on such matters as minimum living and working conditions, anti-discrimination, recognition of the rights of indigenous minorities, environmental standards and protection of biodiversity.

Some core elements in this transition, as they relate to Australia's tropical savannas, are outlined in Table 1. This table provides the basis for much of the subsequent discussion. The table should not be interpreted as implying that industrial values and goals have been wholly displaced by postindustrial values and goals. More appropriately, the evidence suggests that industrial and post-industrial values and goals are now often in contention with each other in shaping resource-related policies and outcomes, with industrial values frequently prevailing where commodity values are high, and post-industrial values prevailing in almost all other contexts. Given that commodity values are very low or negligible over most of the tropical savannas, accordingly postindustrial values will prevail, more especially in those many contexts where savanna resources are well suited to meeting these values. Indeed, the lack of success of the northern tropical savanna zone in serving commodity values may well enhance its national role in serving post-industrial amenity values.

3. IMPLICATIONS FOR THE TROPICAL SAVANNA ZONE

As recently as two decades ago, resource values and land-use options for the tropical savannas were limited and founded upon self-evident truths. As the only broad-acres use capable of generating a significant income stream, pastoralism was pre-eminent, given near-absolute priority to occupy any land deemed suitable and capable of being displaced only by localised more intensive uses, such as mining, agriculture or urban development. Land unwanted for pastoralism would remain vacant or be set aside for Aboriginal reserves. Only negligible allocations had been made for National Parks or other conservation or heritage purposes. Aboriginal land rights were not yet a major public issue and had yet to achieve any legislated or judicial recognition.

At that time pastoralism offered high promise of dramatic productivity gains, focusing mainly on new pasture technologies, supported by improved livestock breeding, supplementary feeding and greatly improved transport systems. Outcomes have fallen far below expectations, even on the core pastoral lands, where productivity gains have been modest and marred by agronomic and economic problems (McCosker and Emerson 1982, McCosker et al. 1988) and by mounting evidence of ecological unsustainability (Pressland and McKeon 1989, Williams 1990). On the extensive tracts of submarginal lands, the structural weakness of pastoralism has become evident, most strikingly revealed by the incapacity of marginal producers to satisfy the minimum standards for livestock management required by the national Brucellosis and Tuberculosis Eradication Campaign (Holmes 1990). Some national leaders in the cattle industry have acknowledged that the national prospects for the industry could be enhanced by a retreat of pastoralism from these submarginal lands.

At the same time, the former hegemony of pastoralism is increasingly questioned, with the proliferation of alternative resource values and land-use options. Many of these new values are closely tied to recently identified major national goals, which are increasingly influential in reshaping resource use. These include: preservation of biodiversity; ecologically sustainable development; new modes of non-consumptive or low-impact resource use; and pursuit of social justice, with emphasis on recognising the rights and needs of Aboriginal peoples. On this latter issue, the northern savannas are already playing a major role in soothing the national conscience, as evidenced by the high concentration of Aboriginal land tenures in the northern savanna zone, sharply revealed in the 1993 AUSLIG map of Australian land tenure (see Figure 1 in Holmes, Chapter 3).

In addition to these nationally articulated goals, another major post-industrial trend is growth in recreation and tourism, as well as in environmental and residential values. Already tourism yields a higher regional income than does the cattle industry (Harris 1992) and there is potential for steady long-term growth.

Closely linked to this is the increase in discretionary lifestyle values which is most clearly expressed in 'sunbelt' migration. While the prime destinations have been along the east coast, as far north as Port Douglas, there are prospects for spill-over into other attractive and accessible parts of the tropical savannas. Even if numerically small, these newcomers could have significant local impacts.

The general trend, then, is to place increasing value on natural resources which can directly satisfy needs and wants through personal involvement, whether by Aboriginals or other peoples, and also to serve newly prominent national goals, rather than produce material goods. The tropical savannas have had little success in producing material goods, other than minerals: they show greater promise in meeting these new goals.

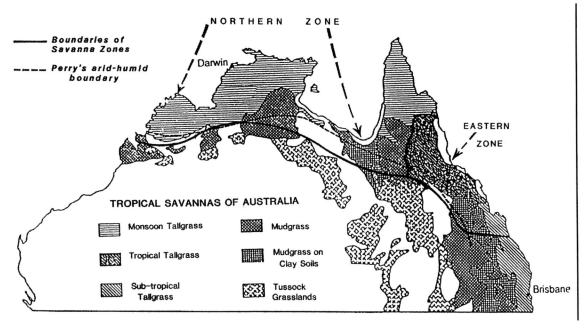

Figure 1. Main vegetation types in the tropical savannas of northern Australia (from Mott *et al.* 1985).

3.1 Location and General Description

The tropical savannas encompass almost all the coastal and subcoastal lands of northern and eastern Australia, north of the Tropic of Capricorn. While areas with a similar physiognomy occur south of the tropic and are defined as Subtropical Tallgrass Savannas by Mott *et al.* (1985), these subtropical savannas have closer affinities to southern agricultural regions in their agricultural potential, as well as in current levels of land use and land management. The incidence of winter rains means that there is a higher dependence on winter crops, while production levels are enhanced by extensive areas of fertile clay soils (vertisols) and by proximity to major population centres.

For the tropical savannas, the boundary is placed generally slightly inland of Perry's (1968) boundary for economic cropping and sown pastures, which he used to define the outer boundary of Australia's arid rangelands (Figure 1).

In the ensuing discussion, the northern savanna zone, dominated by Monsoon Tallgrass Savanna is differentiated from the eastern zone dominated by Tropical Tallgrass Savanna. The two zones, shown in Figure 1, differ in their current levels of rural development and land use. These differences are only partly attributable to differences in pastoral potential but are more influenced by contrasts in accessibility and regional infrastructure (Holmes and Mott 1993).

These environmental and locational differences have led to land-use outcomes which have become more pronounced in recent decades, as well as being of increasing significance in setting future directions. In the eastern zone higher livestock densities and turnoff rates, together with lower input costs, have provided economic incentives to investment in breeding, feeding and general herd management. Almost the entire zone remains securely held in pastoral use, albeit with growing concerns about land degradation, whereas the northern savannas zone comprises mainly marginal or submarginal pastoral lands with a high, increasing proportion in non-pastoral tenures. These zonal contrasts are most clearly revealed in maps of livestock density and land tenure (Figure 2 and also Figure 1 in Chapter 3).This differentiation into two zones provides a useful, if crude, basis for further discussion of changing resource values. Inevitably, the main focus is on the northern zone, not only because of its larger area, but also because it is the zone of most significant resource revaluation. Within this zone an emerging differentiation into three distinct regional categories can be discerned, namely: core pastoral regions where grazing use will remain dominant and institutional changes will be only modest; marginal or 'frontier' regions of diverse resource values, whose recognition and effective use requires significant institutional reform; and localised, urban-oriented development regions which present a distinct set of challenges in managing land-use change (see Chapter 3).

4. Reappraisals of Tropical Savanna Resources

In recognition of changing resource values, new approaches to land-resource appraisal are now emerging, again with the tropical savannas at the forefront in this new research thrust. Just as the tropical

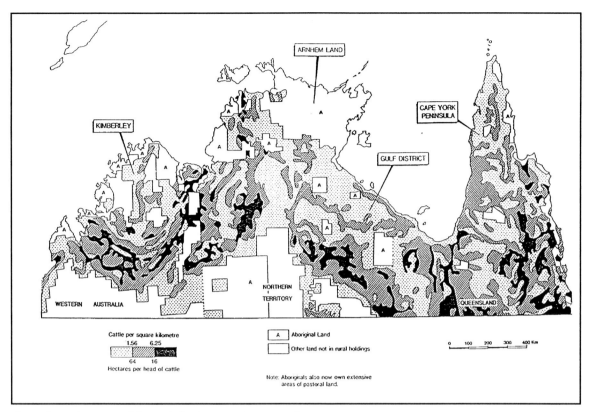

Figure 2. Land tenure and livestock density in northern Australia.

savannas presented challenges which led to innovative methods of land systems classification around mid-century, so also new challenges are spurring a fresh wave of innovation.

In response to prevailing commodity-oriented values, the pioneering CSIRO land systems surveys in the 1940s and 1950s were undertaken in response to a development-oriented agenda, significantly with an initial strong focus on the tropical savannas, in recognition of their promise as a development frontier. With a classification and mapping methodology and output which focused on inherent, land-related attributes, capable of serving many different purposes, these surveys can claim to be in the tradition of pure research even if within an applied context. Their prime purpose was in identifying opportunities for agricultural and pastoral development, perhaps involving more focused land capability surveys. Nevertheless, their continuing value for a variety of purposes has been amply demonstrated. See, for example, their useful role as a basis for viability assessment of marginal grazing leases in the Northern Territory Gulf District (Holmes 1990).

Over the ensuing decades, northern research can be characterised as a purposeful, productive mix of pure and applied research, inevitably with a continuing commodity orientation, with the main outcomes being enhanced knowledge of natural systems (mainly pure research) and of responses to human intervention (mainly applied research). Consistent with changing national priorities, there has recently been a shift towards urgently pursuing a more diverse set of research goals, while also incorporating advances in scientific knowledge and information technologies. More effort is being directed towards specialised surveys and inventories, using the new powerful tools available under computerised Geographic Information Systems (GIS), in expectation of integrated interactive analysis and modelling of the dynamics of environmental responses to natural or human-induced perturbations.

At the same time there remains a continuing need for land classifications as a basis for capability and suitability appraisals for a diverse set of actual or potential uses, and with more attention to appraisal of sensitivity and resilience, in order to assess environmental outcomes. We can expect to see more capability/suitability surveys focusing on: Aboriginal resource management and use; protection of biodiversity and of target species and habitats; recreation and tourism of various types and intensities; and defence needs. Already some land appraisals for these purposes have been undertaken. Since land capability and related issues of carrying capacity and sustainability are addressed by Winter and Williams (see Chapter 2), and are further considered in other papers with specific reference to grazing, Aboriginal use, tourism and national parks, there is no need to discuss these issues here.

However, some comments on the changing geography of land-related resource values in the tropical savannas can appropriately be made, relevant to research priorities, as well as to policy directions and institutional reform. In lands of low productivity, such as the arid and the tropical savanna marginal lands, grazing pressures are concentrated on limited, resource-rich locales, with some dispersal mainly in good seasons. The newly emerging resource values are also highly concentrated at these locales, whether they be Aboriginal subsistence and spiritual values, private recreation, commercial tourism, or protection of biodiversity. Problems of overuse, deterioration, congestion and conflict will be compounded. Not only will there be a propensity for one class of user to overuse the resource, whether it be recreationists or Aboriginal subsistence activities, but there will be competition and conflict between different classes of users. Efforts directed towards effective management for multiple use will often be confounded by a high level of incompatibility between different classes of users. There are formidable problems, for example, in reconciling Aboriginal traditional but technology-assisted uses with traditional (i.e. characteristically) Australian modes of unmanaged, outback recreation activities, and these, again, are hardly compatible with protection of environmental values.

These emerging pressures, incompatibilities and conflicts add urgency to the task of pursuing new directions in resource surveys, assessing values, sensitivity, resilience, carrying capacities, capabilities and suitabilities for various prospective alternative uses, of the type now being explored by various state and territory agencies.

The need for such surveys is well illustrated by considering the growing demands created by a highly diverse array of recreational and tourist activities, now generating pressures, even though intensity of use is still relatively low. A significant problem arises simply because the tropical savanna region does not offer high-quality beachfront recreation. Australia's east, west and, to a lesser extent, south coasts are blessed with prime beachfront resources of high recreational diversity and very high resilience and carrying capacity which admirably serve to reduce recreational pressures on other environments. However, any brief beachfront experience along most of Australia's northern coast will soon dissipate expectations of quality recreational experience. Instead, the north's attractions lie with its imposing array of relatively unspoilt estuaries, rivers, waterfalls, pools, gorges, wetlands, escarpments, karst ridges, lost-city formations, Aboriginal heritage sites, and contemporary Aboriginal culture. Natural resources values are highly localised at attractive but confined sites often of high sensitivity and low resilience, leading to possible widespread replication of the 'Macdonnell Range syndrome' of excess visitor pressure focused on a small set of fragile, confined sites, where negative outcomes are being only partly mitigated by management.

An exceptionally high proportion of recreational values are attached to the tropical savanna's waterways, comprising a diverse mix of estuaries, perennial streams, wetlands, waterfalls, pools and gorges, generally in near-pristine condition and fringed with gallery rainforests, paperbarks, eucalypts or mangroves. Rivers such as the Prince Regent, Drysdale, Ord, Victoria, Daly, Katherine, South Alligator, Roper, Limmen Bight, Macarthur, Calvert, Gregory, Gilbert, Einasleigh, Mitchell, Holroyd, Archer, Wenlock, Jardine, Normanby, Herbert and Burdekin will all act as magnets for an increasingly numerous and diverse band of recreationists, while gorges such as Prince Regent, Geikie, Ord, Katherine, Glyde and Lawn Hill all rank high in scenic values. With such a large array of near-pristine riverine systems, the tropical savannas are exceptionally well endowed when compared to the rest of mainland Australia. Nevertheless, a full inventory of these streams together with an appraisal of prospective uses, users and outcomes will surely reveal that the traditional Australian approach, based on open access and non-management, will soon become impractical. Initially this will occur at localised pressure spots but progressively will expand through these riverine/estuarine environments, with the most observable outcomes on riparian lands and ecosystems but with comparable, if less visible, damage to the streams.

A variety of strategies will need to be adopted, including, for example, regulations on access to and use of streams and riparian lands, with these being related to the tenure of adjacent land. At a later stage, there will be a need to designate areas for specific uses, or non-use. This will require a more coordinated, scientific appraisal of riverine/riparian systems, well beyond the very restricted special-purpose surveys so far undertaken. For example, given the very substantial research agenda encompassed in the Natural Resources Analysis Programme within the Cape York Peninsula Land Use Strategy, it is surprising to find no project directly concerned with riverine/riparian systems, and only two indirectly related, namely Wetland Fauna Survey and Fish Fauna Survey (Freshwater and Estuarine). This gap may be indicative more of current research biases within the scientific community than of the research needs for an effective land use strategy for the peninsula. Riparian land management is now a central component in environmental research on the federal lands of the United States. Given the central importance of river systems to the natural values of the peninsula, a comparable focus must surely emerge.

Given that issues of land tenure, property rights and strategic regional planning are covered in another paper, it would be redundant to pursue them here, other than to re-emphasize needed directional changes in scientific research, as well as a gradualist approach towards strategic planning and multiple resource management, learning from practical experience in high-priority pressured locales, such as Kakadu National Park, as guidelines for the future.

Table 2: Australia's Northern Tropical Savanna Zone: 1991 Population by Location and Aboriginality

Location	Aboriginal or Torres Strait Islander	Other	Not Stated	Total
Numbers				
Urban (>1000)	15305	86066	3986	105357
Rural Locality (200 - 999)	11531	4482	507	16520
Rural (< 200)	10099	11840	829	22768
Total	36935	102388	5322	144645
Percentage Location by Composition				
Urban	41.4	84.1	74.9	72.8
Rural Locality	31.2	4.4	9.5	11.4
Rural	27.3	11.6	15.6	15.7
Total	100	100	100	100
Percentage Composition by Location				
Urban	14.5	81.7	3.8	100
Rural Locality	69.8	27.1	3.1	100
Rural	44.4	52.0	3.6	100
Total	25.5	70.8	3.7	100

5. POPULATION COMPOSITION AND DISTRIBUTION

Compared with all other Australian coastal, non-arid zones, the tropical savannas are distinctive in their very small population numbers, at very low densities and with a uniquely high proportion of Aboriginal peoples.

With 15.4% of Australia's area, the northern and eastern tropical savanna zones contain only 0.8% of Australia's population. Extending over an area of 846 592 km^2, the northern zone contained 144 645 persons in 1991 at an average density of 0.16 per km^2. If the narrow coastal strip is excluded, Queensland's eastern savanna zone contained only 54 600 people at an average density of 0.18 per km^2. In contrast, the adjacent narrow coastal zone, with a mix of rainforest and savanna biomes, contained over 450 000 persons at an average density exceeding 8 persons per km^2.

Over 25% of the northern zone's population is Aboriginal and Torres Strait Islander, which is the highest proportion in any Australian bioclimatic zone. The Aboriginal/Torres Strait Islander (TSI) population shows an above-average propensity to reside in rural locations and, most strikingly, in small rural localities of 200–999 people, as shown in Table 2. On the other hand, the aggregate rural component of the non-Aboriginal population, at 16.0%, differs only slightly from the overall Australian proportion of 14.9%.

There is a sharp segregation between Aboriginal and non-Aboriginal peoples, revealed at all scales of analysis. Quite apart from the localised, segregated Aboriginal enclaves and camps attached to mainly non-Aboriginal settlements, there is an overall marked bipolarity in the ethnic composition of settlements, with most being readily classified as either Aboriginal or non-Aboriginal. Towns and rural localities are classified and mapped in Figure 3.

Of the 16 'non-Aboriginal' towns, classed as service, mining or other, only five have Aboriginal/TSI populations exceeding 20%. Three of these, Normanton, Batchelor and Wyndham are below 50%, and only in Borroloola and Thursday Island are there Aboriginal/TSI majorities. These towns are classed as open service towns, not only because of their histories but also because of their current structures and functions.

By contrast, the 26 settlements classed as Aboriginal/TSI are 'closed' settlements created to meet the needs of Aboriginal/TSI peoples. All save eight comprise over 90% Aboriginal/TSI peoples, while in only two, Yirrkala and Bamaga, does the proportion fall below 80%, and only barely so.

Aboriginal/TSI people do not form large population concentrations. The two largest Aboriginal/TSI population centres are Darwin, with 4674 and Thursday Island with 1642 Aboriginal/TSI peoples. The largest Aboriginal 'town' is Port Keats with 1233 Aboriginal people.

By contrast, the non-Aboriginal population is highly urbanised and strongly concentrated in Darwin. Including its two satellites, Palmerston and Coonawarra, Darwin's 1991 population of 76 125 comprised 52.6 % of all people in the northern savanna zone. Its attractiveness to non-Aboriginal peoples is exceptionally high, with 66.2% living in Darwin. The next largest towns are Katherine, with 7065 and Kununurra with 4062 people.

The dispersed rural population is very small and declining, numbering less than 23 000 in 1991, down from 26 000 in 1986. Less than 3000 persons are employed in pastoral and agricultural production. The cattle industry has only modest labour needs, with over 1000 head of cattle for each equivalent full-time worker, including casual labour.

As in other frontier regions in western societies, the fragmentation and separatism of remote populations, imposed by physical distance, is reinforced by marked differences in the social, economic, cultural and ethnic characteristics of these very small populations. The sharpest contrasts are those between members of the dominant, national Western society and the indigenous people whose cultural identity is under threat. The sparsely settled peripheries have offered the last refuges of partially-assimilated Indians, Inuit, Lapps and Australian Aborigines, whose cultural separatism poses a multitude of seemingly insoluble policy challenges to national governments, with the conflicting goals of assimilation versus separatism being reflected in programmes for social and economic betterment, economic autonomy, resource control and land rights. It is this dualism which led Rowley (1971) to describe our sparselands as 'colonial Australia' within which many white Australians would undertake limited-term tours of duty in the manner of colonial administrators.

This impermanence of much of the population still persists, militating strongly against social, political and cultural cohesion. A case in point is the fragmentation of Northern Territory communities: ethnically; culturally; administratively; economically; socially; and locationally. Statehood would require not only growth in economy and population, but also the moulding of these disparate, largely transient groups of miners, station-managers, stockmen, construction workers, teachers, public servants and others into one cohesive society. Self-government has contributed towards higher social cohesion and reduced transience for the non-Aboriginal population of the Northern Territory, with Darwin and to a lesser extent, Alice Springs, increasingly acting as powerful forces towards residential permanence and territorial cohesiveness.

On the other hand, Cape York Peninsula and particularly the very remote Kimberley District are still regarded as 'colonial' outposts within their respective states. Crough and Christopherson (1993) have argued that, although comprising only 45% of the Kimberley's population, Aboriginals provide '*the long term demographic base for the region*', being much more likely than non-Aboriginal people to be either at the same address, or within the same SLA in 1991 as they were five years previously (83% compared with 35%).

6. POPULATION AS A SOURCE OF ECONOMIC GROWTH

Although supporting relatively few people, much of the savanna's economic activity is, nevertheless, people-oriented. More so than elsewhere in Australia, regional income is generated in providing services to local people, with a very high dependence on public funding. This is the case in both Aboriginal and non-Aboriginal sectors. In the Northern Territory, the public sector, 'broadly defined' employs nearly 40% of the workforce (Caldwell 1985). Consistent with post-industrial trends, a growing proportion of public funds is for people-oriented programmes such as education, training, counselling, welfare, health and housing. Therefore, there are strong population-induced multiplier effects to peripheral regional economies. These are in contrast to the very low regional multipliers obtained from new modes of commodity-based resource use, such as mining, which increasingly rely upon long distance linkages for labour, management, capital, and material inputs and are almost totally disconnected from local economies (Brealey *et al.* 1988). The increasing use of fly-in-fly-out methods in providing management and labour is one further indicator that industrial-era development based upon commodity production offers diminishing prospects for economic and population growth in remote regions which lack the capacity to service such enterprises. Increasingly the benefits from such frontier resource developments flow directly to Perth, Brisbane, Sydney or Melbourne.

In interpreting a large series of input-output tables for various Australian regions, Jensen and West (1983) have argued that greater attention should be given to household consumption-household income linkages, stating that '*local purchases by firms of inputs appear to be considerably less important than the economic effects of purchases by their employees*'. Noting this effect in the Northern Territory, Mules (1985) suggests that the current policy of the Territory government in encouraging the local manufacture of producer goods is misguided, the more so because there are greater opportunities for import replacement with consumer goods. Therefore, while people continue to be a scarce resource in the north, they are assuming an increasingly important role as the underpinning to regional economies. Accordingly, regional economic forecasts must be closely linked to regional population projections.

Certainly Aboriginal peoples will provide a steady source of population growth. A critical question is whether this growth is absorbed and replicated in remote settlements, conforming in size to the census definition of 'rural locality' of 200–999 persons, or whether these settlements steadily grow into large (and more dysfunctional) 'urban' centres, or whether population disperses

towards homeland outstations or migrates into dominantly non-Aboriginal urban centres. As described later, the size and type of Aboriginal residential settlement is a critical indicator of future Aboriginal cultural pathways. While the future trend in Aboriginal numbers is fairly predictable their distribution is highly unpredictable, and the reverse is the case for the non-Aboriginal population. The dominant locations for non-Aboriginal growth can reasonably be forecast but growth trends, historically volatile, are likely to continue so. This volatility is tied to a high level of residential mobility and to the sharp swings in the economic fortunes of the north, which still remains overly dependent upon government budgetary allocations. In turn, these are the outcome of a complex mix of ever-changing public policies, mainly at federal level, but also within the states and territory governments.

In contrast to the north's reduced capacity to attract population through commodity-production is its enhanced capacity to attract people for amenity and lifestyle reasons, either as visitors or residents. For Australia's subtropical and tropical coastal regions there has been a striking reworking of the image of suitability for European settlement, held earlier in this century. In recent decades, the image of the tropics and subtropics has undergone a remarkable transformation, as work, leisure and general living conditions have become progressively more attuned to a warm climate. Critical to this change in climatic preferences have been both technological advance and changing lifestyles. New technologies have largely eliminated the more arduous manual tasks, but, more importantly, refrigeration, air-conditioning and similar innovations have so ameliorated living conditions that people are switching their preferences away from cool towards warm, or even hot, climates. This reversal has been promoted by new, more leisurely lifestyles, with emphasis on year-round outdoor living with prime attention to sunshine and to water-based recreation: swimming pools; waterslides; surfing; sunbathing; boating; sailing; sailboarding; fishing and so on. Population outcomes for Australia's north have been discussed by Holmes (1988a) and Bell (1992).

While most of this growth can be expected to be concentrated along the subtropical and tropical east coast as far north as Port Douglas, modest population spillover to select favoured tropical savanna locations can be expected, notably around Broome, Darwin and Cooktown. Although small in numbers, they may well have significant impacts on local economies and environments.

It is this nascent capacity of the Australian tropics to attract migrants for lifestyle reasons which critically differentiates our northern frontier zone from the cold northern frontiers of other western nations. While the blessings of modern technology can ease the hardships of life in cold climates, they are unlikely to transform subarctic lands into destinations for discretionary migrants, save for a few hardy souls.

Because of its unique combination of a tropical climate within a reasonably affluent, post-industrial western nation, Australia's north can be expected to pursue its own distinctive course in regional development in coming decades. This course will primarily be determined by the discretionary choices of people and not by the output of material goods.

6.1 Human Settlements in the Tropical Savannas

Human settlements in sparsely populated regions have a number of distinctive characteristics. Those commonly found in Australia's remote settlements are discussed in Holmes (1988c). Remote settlements whose primary function is to act as service centres always have a high degree of centrality (complexity in service provision) relative to size, with a low demand threshold for entry of a service, but with reduced quality and higher costs as a trade-off (Jensen 1983). However, in Australia's tropical savannas, rural populations are so thinly spread and so little reliant upon proximity to towns, that the customary hierarchy of service towns, found even in Queensland's inland pastoral zone, is almost entirely missing, save only for some remnants from earlier mining days, such as Coen, Laura, Einasleigh, Georgetown, Croydon, Normanton and Halls Creek. Populations of these settlements are often minuscule unless boosted by an Aboriginal influx in recent decades.

While considerable research has been undertaken into small remote towns, until recently the focus has been almost entirely on new mining towns (Brealey 1972, Brealey and Newton 1978, Brealey et al. 1988). More recently a broader, comparative perspective has been attempted (Loveday 1982a, Holmes 1988c, Loveday and Webb 1989) while there is a growing focus on Aboriginal settlements (Young 1981, Loveday 1982a, 1982b, Ellanna et al. 1989).

In the near-absence of the normal urban service hierarchy, the settlement system is dominated by highly differentiated, single-function towns, which often spring up in response to localised economic opportunities, or Aboriginal servicing requirements, and can equally rapidly decline when no longer required. This high incidence of specialised settlements is shown in Figure 3. Specialised settlements serving the needs of Aboriginal peoples are numerically dominant, with a very high frequency in the census-defined 'rural locality' category, having between 200–999 persons.

7. FUNCTIONAL SPECIALISATION

Functional specialisation contributes to a sharp differentiation of settlements over a wider array of economic, social and other structural attributes. Differentiation is so pronounced that few generalisations can

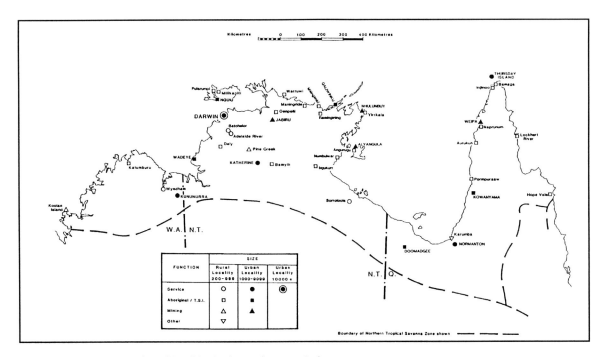

Figure 3. Urban centre and rural localities in the northern tropical savanna zone.

be made about remote towns. Within Australia there has been a very strong research bias towards studying only new mining towns, to a degree where conferences on remote settlements have been concerned only with this special group, under the financial sponsorship of mining companies. While this research has yielded valuable insights into the attributes of these settlements and of their residents, such insights need to be used selectively. The sharp differentiation of remote settlements is well revealed by a summary comparison of the two most common types, the new mining towns and the often new Aboriginal towns. The mining towns are modern transplanted outposts of affluent materialist, Australian suburbia, whereas Aboriginal towns are an expedient, white-imposed means of providing shelter, food and other services to dispossessed, partly detribalised, formerly nomadic hunter-gatherers, having difficulty in reconciling traditional beliefs and customs with the overwhelming onslaught of the dominant materialist culture. Differences are being heightened by recent developments. At one extreme is the advent of the commuter mining resort-type settlements for transients whose home is in a distant city; at the other extreme is the appearance of small Aboriginal out-stations or homeland centres. Aboriginals form these settlements in an effort to remove themselves from Western society and return to a more traditional way of life, thereby making *'this vastness of Australia a human place, as opposed to simply a quarry of some sort'* (Willmot 1984).

A search for commonalities between these two settlement types does reveal a few, emanating mainly from their remote location. These commonalities can generally be construed as divisive rather than unifying factors. They include: exceptionally high levels of economic specialisation; high dependence on remote, powerful decision-makers; minimal levels of community decision-making and of local government structures (though gradually being changed through the fostering of self-management); and high residential mobility and low community identification by transient, specialist service workers undertaking their tours of duty. Both are created as closed towns, with the mining towns being almost wholly company-owned and managed, and with Aboriginal towns being closed to outsiders without permits, save for towns located on or near well-travelled public roads. Even then access for outsiders is only fleeting and superficial.

It is hardly surprising that the tropical savanna population is characterised by a high level of fragmentation and separatism of social groups. This is an outcome of big distances, small numbers and inadequate transport systems, reinforced by the marked specialisation and differentiation of settlements, described above and further accentuated by strong institutional barriers against open access. Apart from the access barriers in mining and Aboriginal towns, already mentioned, there are comparable problems of access to large cattle stations, whose size and self-sufficiency act as a suppressant to local urban growth and impede the development of an open infrastructure needed to serve non-station consumers, including an increasingly large number of travellers (Holmes 1984).

7.1 Aboriginal Settlements

Contemporary Aboriginal society is markedly differentiated along the basic dimension of separate development versus assimilation into the dominant white society. Relevant attributes include degree of ethnic admixture and cultural identity, as evidenced by language, laws, ritual, tribal structure, homeland identification and economic activities. These differences are usually closely tied to geographical location and the existing settlement system, which provides the context either for self-determination or for continuing detribalisation and submergence within white society as an underprivileged subculture.

Only in the arid interior and tropical savannas have Aboriginal groups been able to maintain significant ties to their land, usually on the least productive margins, as revealed on maps of land tenure (Figure 1, Chapter 3), as well as human settlement (Figure 2 in this paper).

While all Aboriginal settlements have small populations by conventional standards, the Aborigines themselves are accustomed to living in very small groups, with some taboos on contacts and with inter-group friction. Large settlements create an environment in which it is difficult to sustain traditional customs and values. There is a loss of contact with traditional lands, a reduction in customary hunting and collecting of 'bush tucker', an excessive amount of alcoholism, petrol-sniffing and other social disorders, and great difficulty in inducting the young into traditional ways.

Recently, in reaction to these social ills, there has been a limited counter-movement towards population dispersion in small out-stations (or homeland centres) comprising a dozen or so persons living on their traditional lands. This movement has beneficial effects in enabling a return to traditional life-styles, with improved health, a reduced dependence on Western technology and controlled access to alcohol. However, it poses severe logistical problems in providing needed services. This problem is only partially mitigated by the continuing close ties to families and services in the main settlement, involving frequent movement of persons and supplies (Loveday 1982a, 1982b).

Further population dispersion may occur through the acquisition of additional lands for Aborigines, including vacant crown lands and various pastoral leases. Physical remoteness is accentuated by entry restrictions on non-Aborigines. This separatism is regarded as vital in allowing Aboriginal groups to renew their search for an Aboriginal identity.

By conventional Western standards, Aboriginal settlements can be classed as residential and service-delivery towns, lacking any substantial economic base, other than from public funding. The delivery of specialist services, including administrators, teachers, store managers, mechanics, nurses and so on, is still mainly in the hands of transient non-Aborigines, with only limited replacement at lower levels of skill and responsibility, even in self-managed communities. Issues relating to underemployment, unemployment, lack of motivation for regular employment, conflicts between traditional obligations and Western ideas of work and wealth accumulation, have received considerable attention (Altman and Nieuwenhuysen 1979, Turnbull 1980, Altman 1983, Young 1981, Fisk 1985, Commonwealth of Australia 1985). The transition from paternalistic control by missionaries or government 'protectors' towards Aboriginal self-management does offer possibilities for the appearance of new modes of economic organisation which reconcile Aboriginal values with the demands of a modern cash-based economy. Stanley (1985) has studied economic change in self-managed communities. While his conclusions are generally pessimistic, he does observe that Aboriginals show a capacity to undertake group savings and investment, where individuals are unable to do so. Both Stanley (1985) and Young (1981) recognise that remote location and limited resource base, quite apart from any cultural barriers, will continue to ensure that Aboriginal settlements remain economically dependent, even though the legal framework is now in place to enable Aboriginal communities to have a strong voice in shaping their own future.

7.2 Mining towns

At the opposite end of the economic spectrum, as in most other respects, are mining towns. In marked contrast to Aboriginal settlements, the rise and fall of mining towns is strictly tied to considerations of supply and demand, and, increasingly, to methods of optimising the provision of human input into mining ventures, as mining enterprises seek alternatives to the conventional on-site mining town.

The maturing of the first generation of modern mining towns has posed a series of major research issues. The most obvious has been the need for further surveys to assess changes in socio-demographic structure, to ascertain the extent to which towns are achieving a stable workforce and population, comparable to such long-established mining towns as Broken Hill, Kalgoorlie and Mt Isa. Also questions of economic diversification and a more varied mix of employment opportunities start to loom large as the numbers of school-leavers continue to grow. Equally important are questions about 'normalisation', the process of conversion into open towns, with the two most important issues being property ownership and local governance.

Companies are increasingly attracted towards normalisation, thereby divesting themselves of ongoing responsibilities in ventures regarded as ancillary to their mining operations. Resident home-ownership has been increasingly fostered, as a means of promoting workforce stability, with a further incentive in the recent loss of the tax avoidance advantages of subsidised housing. The

handover of municipal responsibilities is also becoming increasingly favoured (Carly 1977).

As the pace of normalisation has quickened it has become subject to increasing political controversy. Critics point out that corporations and State governments dominate negotiations, to the disadvantage of communities and local government with companies seemingly willing to 'hire' shires to run the towns and deflect criticisms from themselves (Thompson 1981). In the Northern Territory, normalisation is increasingly seen as a means of pursuing political goals of greater territorial autonomy and of undermining Federal government restrictions on the functions, size and regional impact of Jabiru, the uranium mining town in the Kakadu National Park, whose functions and growth are restrained by national concerns about uranium mining, environmental preservation and Aboriginal land rights (Lea and Zehner 1986).

In terms of regional impact assessment, remote mining towns have long been seen as potential catalysts for economic growth and diversification in remote regions. Results have been well below expectations, and, indeed, impacts have been so slight that there is little to be researched (Linge 1980, Harman and Head 1982). Input-output studies reveal very low cross-sectoral multipliers for mining ventures in remote locations (Jensen and West 1983, Mules 1985).

Social impact assessment has become a matter of increasing attention for almost all major ventures in remote locations, with the most ambitious of these being the wide-ranging East Kimberley Impact Assessment Project, designed to examine the impact of tourism, mining and other developments on the Aboriginal communities, and the response options available to governments, developers and the communities themselves (Coombs et al. 1989).

The single-enterprise mining town of the 1960s and early 1970s has come increasingly under scrutiny, with two major alternatives being canvassed, namely centralised, multi-enterprise towns with some local commuting to worksites, and fly-in fly-out workforces, spending from 7–17 days in extended work rosters before flying out to distant homes for rest and recreation. The feasibility of the fly-in approach was initially shown by employment on oil-rigs and by the comparable land-based Moomba oil-gas venture in South Australia as early as 1968. Recent ventures include Narbalek, Northern Territory (uranium) in 1980, Argyle, Western Australia (diamonds) in 1983 and Kidston, Queensland (gold) in 1985. One major outcome is a further reduction in the regional economic impact of mining, with reinforcement of links to distant, mainly metropolitan centres, up to 2200 km away for the worker commuter from Argyle to Perth.

7.3 Dispersed Rural Settlement

The proportion of the population in the northern savanna zone located in dispersed rural settlements is not much higher than the Australian average and numerically it is only one-third of the population of Darwin (Table 2). The non-Aboriginal rural population is shown to be fifty percent higher than the Aboriginal. However, it is difficult to establish a secure figure for the Aboriginal population, given the high rate of movement between Aboriginal 'towns' (classed mainly as rural localities in the census) and outstations.

A major policy issue arises from the need to provide adequate services to the dispersed rural population. Logistic problems in servicing widely dispersed rural populations are reinforced by the steady decline in self-sufficiency, with new technologies allowing enhanced servicing but with mounting costs. Substantial cross-subsidies are required suggesting the need to develop a more integrated policy approach to facilitate service delivery. The matter becomes more urgent given the current policy thrust towards cost-recovery on services, linked to a wider move towards deregulation and privatisation (Holmes 1988a, 1988b). The relative merits of nucleated compared with dispersed rural settlement as adaptive responses to the effects of isolation within Australia's rangelands are discussed in Holmes (1984).

While all of these issues have relevance in any discussion of dispersed rural settlement in the tropical savannas, one of the most important emerging policy issues concerns the prospective restructuring of rural settlement in response to changing resource values. Within rangelands generally, there is a move away from broad-acres income streams tied to land title and towards other resource values, from which income streams are mainly captured within urban or resort settings. An increasingly important policy issue is to encourage an appropriate mix of settlement types to accommodate these new sources of economic activity. Wherever appropriate, preference should be given to encouraging the growth of new or existing small settlements to serve as open, diversified tourist and service centres, integrated with the local rural economy. This could well be at such pivotal locations as Fitzroy Crossing, Gibb River, Kununurra, Jabiru, Borroloola, Burketown, Karumba, Bamaga, Portland Roads, Coen and Starcke. Every effort should be made to support Aboriginal initiatives in urban/resort-based enterprises, often linked to activities on adjacent Aboriginal lands.

Active encouragement of these 'open' settlements does not necessarily imply discouragement of stand-alone, single-ownership tourist resorts. These resorts can be an adaptive response, particularly in remote, ill-serviced locations and also where sensitive environmental values can best be preserved within the controlled context provided by a stand-alone resort.

In the emerging context of strategic land-use planning for northern peripheral regions, there is a clear need to

link land tenure and land use strategies to flexible, adaptive strategies which foster the growth of settlement systems, compatible with their environs.

7.4 The Role of Major Regional Centres

Any study of human settlement in the tropical savannas would be incomplete without reference to the increasingly pivotal role of major urban centres in regional transformation, in many different ways. Only two are mentioned here. The first is in breaking down the barriers of isolation, through providing an increasingly sophisticated set of services to their hinterlands, to which they are linked by efficient road, air, telephone, television, radio and other services. Their second pivotal role is in transforming the image of the north by demonstrating that tropical lifestyles can be highly attractive, within a society in which the growth of population and employment is increasingly tied to discretionary decisions influenced by amenity and environmental values.

Darwin is of critical importance, not merely because of its near-central location within the northern savanna zone and its remoteness from any other major city, but because of its innovative role as the capital of the Northern Territory, giving it political, economic and cultural influence far beyond its size. Although just beyond the boundaries of the tropical savannas, Cairns and Mt Isa are of vital importance to their hinterlands, as is Townsville to the north-east zone.

In addition to their roles in servicing vast hinterlands, these relatively large population centres are having a growing impact on land use, land ownership and land markets within their immediate hinterlands, particularly so with Darwin, where there is an on-going sharp transformation from low-value to high-value land uses. This creates a need for state intervention in planning and implementing land use changes.

8. CONCLUSION: AUSTRALIA'S TROPICAL SAVANNAS WITHIN CHANGING NATIONAL AND GLOBAL CONTEXTS

Australia's tropical savannas are undergoing rapid changes in resource values, land ownership, economic directions and broad societal outcomes, in ways which could not have been anticipated a quarter-century ago. These changes are likely to continue, often taking new directions which cannot be anticipated. The rapid pace of change within modern societies appears to be magnified in Australia's northern savannas, which are highly susceptible to changes in economic, social and political priorities.

One important overall perspective is to engage in an ongoing review of the place of the tropical savannas within Australia's changing economic and social context. Much of the discussion in this paper, and also in my other paper to this symposium, has been strongly focused on this issue, emphasizing the enhanced role of the savannas in satisfying new environment-oriented and human-oriented goals with a diminishing role in producing commodities, other than minerals. Even if commodity markets were to improve, the basic equations in resource valuation seem to have been irreversibly altered.

The second perspective is to position Australia's savannas within the wider global context. Here there are two obvious, highly divergent bases for comparison: either in terms of bioclimatic affinities with other tropical savanna zones or in terms of socioeconomic affinities with other peripheral zones in advanced, western nations.

Notwithstanding environmental affinities, Australia has always been the exceptional case among the tropical savannas in its economic, social and demographic setting. Whereas other tropical savannas, to varying extents, are being subjected to increasing population pressure, generating heavier demands on their capacity to produce food and other agricultural products, Australia's savannas are under decreasing pressure to engage in agricultural and pastoral production, at least for the time being. Some policy attention and research activity continues to be directed towards increasing production and fostering demand in order to realise this agricultural potential, with very mixed results. The resources of Australia's savannas are increasingly being reassessed in terms of non-commodity outputs, including meeting the aspirations and needs of Aboriginal peoples, preserving biodiversity and satisfying recreation, tourism and other lifestyle values.

In searching for useful comparative experience on regional policies and programmes, Australia has to pay attention to current trends in zones with broadly similar socioeconomic contexts, such as the peripheral zones of other advanced western nations, most notably northern Canada, Alaska and northern Scandinavia, even though these zones have markedly different bioclimatic environments. Already considerable comparative research has been undertaken, mainly under the auspices of the North Australia Research Unit, and there is an expanding series of relevant publications (Jull 1991, Jull and Roberts 1991, Jull et al. 1994). These are having a policy impact, mainly at federal level, where there is an ongoing sensitivity to Australia's international profile and obligations on such issues as meeting the needs of indigenous peoples and preserving biodiversity. Much can be learnt from successes and failures in comparable socioeconomic contexts.

Yet, as already pointed out, comparisons with other northern frontiers must also recognise increasingly significant differences, primarily related to the natural environment, which offers opportunities for Australia's northern 'frontier' to satisfy emerging postindustrial values as a basis for steady economic and population growth. These opportunities do not exist in cold frontier zones.

Because of its unique combination of natural and societal attributes, Australia's tropical savannas seem poised to pursue a uniquely different trajectory in regional growth and change, unmatched by any other major region. In shaping policies, programmes and public institutions, and in identifying research priorities, we will need to be increasingly sensitive to this unique blend of nature and society, and more aware of the ever-changing challenges and opportunities presented by our distinctive northern frontier.

Acknowledgments

I have benefitted from discussions with many people involved in northern research, most notably Nuggett Coombs, David Lea, Peter Jull, John Mott and Elspeth Young. This paper has been written while the author was Research Associate in the Environment and Behavior Program, Institute of Behavioral Science, University of Colorado, Boulder.

References

Altman, J.C. (1983). *Aborigines and Mining Royalties in the Northern Territory*. AIAS, Canberra.

Altman, J.C. and Nieuwenhuysen, J. (1979). *The Economic Status of Australian Aborigines*. Cambridge University Press, Cambridge.

AUSLIG (Australian Surveying and Land Information Group) (1993). *Map of Australian Land Tenure*. Government Printer, Canberra.

Australian Bureau of Statistics (1991). *Census update/Australian Bureau of Statistics*. ABS, Canberra.

Bell, M. (1992). *Internal Migration in Australia*. Bureau of Immigration Research, Canberra.

Brealey, T.B. (1972). *Living in Remote Communities in Tropical Australia*; 1. Exploratory study. CSIRO Division of Building Research, Canberra.

Brealey, T.G. and Newton, P.W. (1978). *Living in Remote Communities in Tropical Australia: the Hedland study*. CSIRO, Division of Building Research, Canberra.

Brealey, T.B., Neil, C.C. and Newton, P.W. (eds) (1988). *Resource Communities: Settlement and Workforce Issues*. CSIRO, Melbourne.

Brotchie, J.F., Hall, P. and Newton, P.W. (eds) (1987). *The Spatial Impact of Technological Change*. Croom Helm, London.

Caldwell, P.J. (1985). Public sector in the Northern Territory economy. In: *Economy and People in the North*, (eds Loveday, P. and Wade-Marshall, D.). NARU, Darwin.

Carly, P.J.L. (1977). *Opening Up: A Report on the Considerations Involved in the Progressive Establishment of the Normal Roles of Local Authorities, Government Agencies and the Communities in Mining Towns*. Wickham, Perth (mimeo).

Coombs, H.C., McCann, H. Ross, H. and Williams, N. (1989). *Land of Promises: Aborigines and Development in the East Kimberley*. Centre for Resource & Environmental Studies, ANU and Aboriginal Studies Press, Canberra.

Commonwealth of Australia (1985). *Report of the Committee of Review of Aboriginal Employment and Training Programmes*. AGPS, Canberra.

Crough, G. and Christophersen, C. (1993). *Aboriginal People in the Economy of the Kimberley Region*. NARU, Darwin.

Ellanna, L., Loveday, P., Stanley, O. & Young, E. with assistance from Ian White (1989). *Economic Enterprises in Aboriginal Communities in the Northern Territory*. NARU, Darwin.

Fisk, E.K. (1985). *The Aboriginal Economy in Town and Country*. AIAS, Canberra.

Harman, E.J. and Head, B.W. (eds) (1982). *State, Capital and Resources in the North and West of Australia*. University of WA Press, Perth.

Harris, P. (1992). *An Economic Strategy for Northern Australia*. James Cook University, Townsville.

Holmes, J.H. (1984). Nucleated rural settlement as a response to isolation: the large cattle station. In: *North Australia: The Arenas of Life and Ecosystems on Half a Continent*, (ed Parkes, N.D.). Academic Press, Sydney.

Holmes, J.H. (1988a). New challenges within sparselands: the Australian experience. In: *Land, Water and People: Geographical Essays in Resource Management and Organisation of Space in Australia*, (eds Heathcote, R.L. and Mabbutt, J.A.), pp. 75-101. Academy of Social Sciences, Canberra.

Holmes, J.H. (1988b). Private disinvestment and public investment in Australia's pastoral zone: policy issues. *Geoforum*, **19**: 307-322.

Holmes, J.H. (1988c). Remote settlements. In: *The Australian Experience: Essays in Australian Land Settlement and Resource Management*, (ed Heathcote, R.L.), pp. 68-84. Allen and Unwin, Sydney.

Holmes, J.H. (1990). Ricardo revisited: submarginal land and non-viable cattle enterprises in the Northern Territory Gulf District. *Journal of Rural Studies*, **6**: 45-65.

Holmes, J.H. and Mott, J.J. (1993). Towards the diversified use of Australia's savannas. In: *The World's Savannas: Economic Driving Forces, Ecological Constraints and Policy Options for Sustainable Land Use*, (eds Young, M.D. and Solbrig, O.T.), pp. 283-317. UNESCO, Paris and Parthenon, Carnforth, UK.

Jensen, R.C. (1983). Economic problems facing small towns in the arid zone of Australia. In: *Design for Arid Regions*, (ed Golany, G.). Van Nostrand Reinhold, New York.

Jensen, R.C. and West, G.R. (1983). The nature of Australian regional input-output multipliers. *Prometheus*, **1**: 202-221.

Jull, P. (1991). *The Politics of Northern Frontiers in Australia, Canada and Other 'First World' Countries*. NARU, Darwin.

Jull, P., Mulrennan, M., Sullivan, M., Crough, G. and Lea, D.A.M. (eds) (1994). *Surviving Columbus, Indigenous Peoples, Political Reform and Environmental Management in North Australia*. NARU, Darwin.

Jull, P. and Roberts, S. (eds) (1991). *The Challenge of Northern Regions*. NARU, Darwin.

Lea, J. and Zehner, R. (1986). *Yellowcake and Crocodiles*. Allen and Unwin, Sydney.

Linge, G.J.R. (1980). From vision to pipe dream: yet another northern miss. In: *Of Time and Place*, (eds Jennings, J.N. and Linge, G.J.R.). ANU Press, Canberra.

Loveday, P. (ed) (1982a). *Service Delivery to Remote Communities*. NARU, Darwin.

Loveday, P. (ed) (1982b). *Service Delivery to Outstations*. NARU, Darwin.

Loveday, P. and Webb, A. (eds) (1989). *Small Towns in Northern Australia*. Conference Proceedings, 22-24 February 1989, NARU, Darwin.

Loveday, P. and Wade-Marshall, D. (eds) (1985). *Economy and People in the North*. NARU, Darwin.

McCosker, T.H. and Emerson, C.A. (1982). Failure of legume pasture to improve animal production in the monsoonal dry tropics of Australia: a management view. *Animal Production in Australia*, **14**: 337-340.

McCosker, T.H., O'Rourke, P.K., Edington, A.R. and Doyles, F.W. (1988). Soil and plant relationships with cattle production on a property scale in the monsoonal tallgrass tropics. *Australian Rangeland Journal,* **10**: 18-29.

Mott, J.J., Williams, J., Andrew, M.H. and Gillison, A. (1985). Australian savanna ecosystems. In: *Ecology and Management of the World's Savannas,* (eds Tothill, J.C. and Mott, J.J.), pp. 56-82. Australian Academy of Science, Canberra.

Mules, S.T. (1985). Impact measurement in Northern Territory economic development. In: *Economy and People in the North,* (eds Loveday, P. and Wade-Marshall, D.). NARU, Darwin.

Perry, R.A. (1968). Australia's arid rangelands. *Annals of the Arid Zone,* **7**: 43-9.

Plumb, T. (ed) 1982. *Atlas of Australian Resources.* Volume 3: Agriculture. Division of National Mapping, Canberra.

Pressland, A.J. and McKeon, G.M. (1989). Monitoring animal numbers and pasture condition for drought administration - An approach. *Range monitoring workshop. 5th Australian Soil Conservation Conference,* Working Papers.

Rowley, C.D. (1971). *The Destruction of Aboriginal Society.* Australian National University Press, Canberra.

Stanley, O. (1985). Economic development problems in remote, Aboriginal communities. In: *Economy and People in the North,* (eds Loveday, P. and Wade-Marshall, D.). NARU, Darwin.

Thompson, H.M. (1981). "Normalisation": industrial relations and community control in the Pilbara. *Australian Quarterly,* **53**: 301-324.

Turnbull, S. (1980). *Economic Development of Aboriginal Communities in the Northern Territory.* AGPS, Canberra.

Williams, J. (1990). Search for sustainability: agriculture and its place in the natural ecosystem. In: *Agriculture and the Ecosystem in North Queensland,* (eds Norman, K.L. and Garside, A.L.), pp. 21-42. AIAS Occasional Publication No. 51, Australian Institute of Agricultural Science, Townsville.

Willmot, E. (1984). *Repopulation of remote Australia.* Australian Population Association National Conference, Sydney (mimeo).

Young, E. (1981). *Tribal Communities in Rural Areas.* Aboriginal Component in Australian Society Series, Australian National University, Canberra.

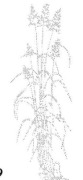

Chapter 2

Managing Resources and Resolving Conflicts: The role of science

W.H. Winter[1]
and J. Williams[2]

[1] CSIRO Division of Tropical Crops & Pastures,
306 Carmody Rd
St Lucia, QLD 4067
Australia

[2] CSIRO Division of Soils,
GPO Box 639
Canberra, ACT 2601
Australia

Abstract

Scientists face the challenge of developing an understanding of savanna ecosystem function from a holistic perspective, while recognising that land use and condition is affected as much by social and economic conditions, and the attitudes of the land users, as it is through the biophysical processes. In addition to the capabilities imposed by natural laws, savannas have values and capabilities inferred by the range of claimants/ users and will therefore require different management strategies to preserve or enhance these values and capabilities. While attempts have been made to place an economic value on particular uses of savanna lands, these analyses have generally ignored the different value systems, the concept of resource rent (as a cost to offset reversible environmental change), or capital charges (as a cost of irreversible environmental change).

There is considerable controversy in the literature and misunderstanding between user groups concerning the utility of indicators of sustainability and degradation, particularly since indicators cover the spectrum of economic, social and biophysical issues. Undoubtedly, appropriate management of the savannas is compromised by these different perspectives and, in many instances, the different time-frames of the interest groups. While scientific knowledge is of fundamental importance, it will be but one of the factors contributing to better management of the tropical savannas. Such information will be an input into a complex social and economic framework, where consensus amongst such diversity may only be possible if there is a shared vision for the future of tropical savannas.

The major areas where science can contribute are savanna ecology, land use impacts on the soil, vegetation, animal and water resource base and a systems analysis approach to integrated management. The current research effort is large, rich and diverse, but fragmented, and in many cases, reactive. Consolidation of this effort may be an undesirable outcome, as one organisation, even located at a few sites, could not adequately address the complexity of ecological issues within this large and diverse ecozone, let alone the interactions with the potential users. Rather, maximal utility of information from this rich and vast array of R&D effort could be obtained from greater interaction, some rationalisation and planned coordination following a collaborative strategic planning process.

1. INTRODUCTION

Tropical savannas are a complex vegetation type, where grasses, shrubs and trees co-exist in dynamic equilibrium. Humans have lived in and exploited the savannas for thousands of years and may, in some cases, have influenced the very structure of the vegetation and the faunal composition. However, the past century has seen great changes in the intensity and type of exploitation as a consequence of colonisation, population growth, industrialisation and developments in transportation and communication. In more recent times, there has been questioning of the balance between agricultural and non-agricultural uses of the savannas at the regional, national and international levels. Managing the savannas for the best outcomes for sequestration of greenhouse gases, surface or subsurface water quality, contribution to biological diversity and off-site effects will continue to assume increasing importance.

The broad challenge facing science is to understand the functioning of the savanna ecosystems from a holistic perspective, so that management options, for any use and from any perspective, can be based on this understanding. Importantly, the interactions of the savannas with other biological and social systems need to be understood to allow balanced decisions to be made and to avoid unforeseen negative outcomes.

Information gathered through the scientific process and integrated with social and political demands of the day will be critical in moving towards sustainable use of the tropical savannas. Such progress will require a radical change in the mind-set by many of the current 'players' in savanna research and development (R&D). Biologists must recognise that land use and condition is affected as much by socio-economic conditions and the attitudes of the land users as it is by the underlying biophysical processes. Current land users will increasingly have to accept the legitimate 'constraints' imposed by the state — particularly if urban taxpayers' money is used either directly or indirectly to support infrastructure development and maintenance and more sustainable management practices. The mining sector, in the main, has accepted this situation and responded positively.

This discussion focuses upon Australian tropical savannas but draws upon relevant information from other regions where appropriate. This concentration is warranted given the unique social and economic circumstances within Australia, such as the extremely low population density, negligible cropping, extensive management practices, large areas managed for recreational or conservation purposes, the large areas of Aboriginal lands, and the modest quality of services and communication within the region.

1.1 Savanna Ecosystem Utility and Conservation

Claimants to savanna use include traditional use by indigenous peoples, agriculture, mining, military, tourism and conservation of biodiversity. Consequently, savannas will have different values and attributed capabilities for each of these claimants/users and the savannas may require different management to preserve or enhance these values and capabilities as perceived by the diversity of users. However, completely apart from the belief systems of users, there is a physical limit to savanna capability which is determined by the initial conditions and the physical laws governing the behaviour of energy and matter. Spiritual significance would, of course, be excluded from this precondition.

The most obvious, and widest use of savannas continues to be various forms of pastoralism. Consequently, the majority of current scientific assessments of savanna land capability have been based upon agricultural use, ranging from the most extensive nomadic pastoralism to the intensive integrated crop and livestock systems. A fundamental tenet of these assessments was that the natural resources of the savannas were to be exploited for short to medium term benefit. In more recent times there has been an increasing recognition of the need to evaluate the multiple land use options for savannas, all with different notions of exploitation, and of the intrinsic and social values of savannas. Not surprisingly, we now find that few scientific studies can provide adequate information on the impacts of non-agricultural or multiple land use upon the structure of the vegetation, biodiversity, ground water recharge, soil dynamics, nutrient and energy flows, responses to climatic variability, and community resilience.

Attempts have been made to place an economic value on particular uses of savanna lands. Many of these evaluations have assumed that there is a safe 'carrying capacity', or exploitation rate (e.g. livestock grazing, vehicular traffic, people pressure, harvest of native plants and animals, withdrawal of ground-water and damming of natural waterways, extraction of soil and minerals). Such analyses have ignored the concept of resource rent, as a cost to offset reversible environmental change, or capital charges as a cost of irreversible environmental change.

A recent analysis of bio-economics of agricultural use in South Africa by Mentis and Seijis (1993) describes how, given their social constraints and imperatives, Black South Africans set livestock densities at the 'ecological carrying capacity', i.e. the vegetation-animal equilibrium that the land can bear, rather than the one third less 'economic carrying capacity' for efficient commercial production. Based upon the relatively steady state of stock numbers over a 10–20 year period they concluded that the ecological carrying capacity is sustainable. Such an analysis is typical of the single-purpose view of savanna value and ignores the fact that livestock numbers is one of the least sensitive indicators of change, that irreversible losses of such attributes as biodiversity, soils and social amenity are likely to have

occurred, that offsite effects which are costs to other groups have occurred and that fluctuations in annual income or standard of living may have become greater. All of these outcomes have a cost, some borne by the current generation and some externalised.

Other examples have recognised and legitimised the social and ecological values of particular savanna use and compared them with the potential economic value. Some cases, such as the recent Coronation Hill case in Australia, where conservation and Aboriginal values were weighed against potential mining income, have been resolved in favour of the former — but not before a long and embittered battle where scientific information was questioned and values tested. Nested within its broader role, science has a unique and invaluable role to play in providing biophysical information and the tools which can be used to conduct analysis for resolution of such conflicts and ultimately for the sustainable exploitation of savannas. The LUPIS software (Ive and Cocks 1988) is one such tool which has been used in the savannas of Cape York Peninsula of northern Australia. The 'WhatIf' software described by Veitch (1992) is another example of a systems analysis package which will have increasing value in resolving conflict over multiple objective resource use.

2. INDICATORS OF SUSTAINABILITY: MEASURES OF LAND AND WATER RESOURCE DEGRADATION

Discussion among groups which share an interest in the management of savannas is not aided by the plethora of indicators of sustainability (Hutchinson 1992) or by the different agendas and objectives of the large number of divergent working groups which have proposed sets of indicators. Governments usually set goals at the broadest level to reflect their purpose for resource use. For instance, the National Strategy for Ecologically Sustainable Development (Ecologically Sustainable Development Steering Committee 1992) does not contain any performance indicators, rather stating that the goal is: *"development that improves the total quality of life, both now and in the future, in a way that maintains the ecological process on which life depends"*. The National Strategy for the Conservation of Australia's Biodiversity (Australian and New Zealand Environment and Conservation Council Task Force on Biological Diversity 1993) goes a step further by setting action priorities and time-frames, but there is no articulated way of determining whether we will be closer to conserving biodiversity in a decade than we are now. By contrast, the Standing Committee on Agriculture and Resource Management (1993) statement on Sustainable Agriculture specifies indicators of

- financial viability of agriculture in a region
- environmental stability as evidenced by on-farm nutrient balance and natural vegetation retention, and
- social factors such as age and skill levels of farmers.

While some may not agree with these performance indicators they are at least an attempt to define some measurable outcomes of policy in terms of economic, environmental and social goals. Conservationists may well set indicators at a much finer scale, with attributes reflecting the biophysical functioning of the systems (e.g. fauna and flora biodiversity, soil health, water quality) while economists will seek more detailed indicators of on-farm economic analysis.

Considering the above, there are two important points to make. Chosen indicators will always reflect the perspective of the person or group setting the performance goals, ie, they are not without bias. Additionally, they should be open to change, particularly to accommodate new knowledge or understanding.

Given this background, it is not surprising that there is considerable controversy and misunderstanding between groups surrounding the issues of indicators of sustainability and degradation, particularly since indicators cover the spectrum of economic, social, biological and physical issues, and quite naturally, the various user groups attribute different weighting to these issues. It is possible that one user group's positive indicator of sustainability may embody attributes which are clearly negative to another user group. Such an example can be found in the semi-arid woodlands of northern Australia where, under certain conditions, e.g. good seasons when feed is not limiting, cattle growth is improved (a positive attribute) when the pastures are 'deteriorated', as characterised by a reduction in ground cover, perennial grasses, soil surface condition and soil organic matter content (negative attributes, Ash *et al.* 1996, McIvor *et al.* 1995). Other examples can be found in Africa where the provision of additional watering points has improved the economic and social sustainability of the herdsmen while the natural grazing lands have undergone substantial biophysical degradation.

The diversity of definitions of degradation reflect those of sustainability outlined above. Unfortunately appropriate management of the savannas is compromised by these differences which, in many instances, operate on different time-scales. A survey of people considered important in the management of savannas in southeast Queensland (MacLeod and Taylor 1994) indicated that time-frames varied from 3–5 years for financiers, 5–10 years for graziers, 50–100 years for scientists and >150 years for conservationists. The importance of the timeframe, and the nature of the ecosystem response to perturbation in terms of stability, resilience and resistance, are defined and set down in some detail by Williams *et al.* (1993). It is important that there is a movement towards agreement in the use of these concepts and terminology. At present, this is not the case. From a natural resource viewpoint degradation is related to the changes in the savannas which make them less capable of meeting their desired uses, which must include the full range of claimants. As set out by

Williams *et al.* (1993), it is important to recognise that there are degrees of degradation, usually considered a continuum from no change to change which is irreversible over any timeframe.

The forms of resource degradation are diverse, ranging from soil loss, biodiversity loss, invasion of woody weeds and feral animals and changes in water quality. Surveys of degradation of the world's savannas are usually incomplete as data have been collected from a limited perspective, but in many cases missing information can be inferred from relationships between components of systems, e.g. floristic changes leading to changes in faunal diversity and composition.

3. THE ROLE OF SCIENCE

Scientific knowledge will always be one of the primary factors contributing to better management of the tropical savannas. Science and technology is not only able to contribute knowledge on the biophysical processes, but also the analytical tools which facilitate this scientific knowledge being integrated into a complex social and economic framework to yield optimum solutions to conflicting land use. It is our view that maximum effectiveness of scientific knowledge will best be developed within this context, without losing site of emerging scientific issues and research opportunities. Importantly, scientists will need to maintain an impartial 'honest broker' role, providing unbiased, sound and broadly based information. It would be naive to believe that individual scientists will be able to maintain an unbiased position. Collectively, however, scientists are probably best placed to provide a balanced view, given that questioning, dispute and public airing of information and opinion is part and parcel of this discipline.

There is potential for this role to be compromised as R&D organisations conduct research for specific savanna users. This conflict will become less of a problem in the future with increasing maturity of the sectors, including mutual recognition of legitimacy. It is also timely for funding agencies to consider their obligations to future generations rather than only to the current interest groups.

The multiple and often competitive uses of tropical savannas must consider the high social and amenity value placed on savannas, management of weeds and feral animals, International obligations (World Heritage areas, Biodiversity Convention, Global Climate Change), and regulatory-based monitoring programs of land condition. Social, economic, biophysical problems and inevitable conflicts will arise from this array of use and expectation of the savanna resource. As mentioned above, science can provide integrative tools which can facilitate the rational analysis of the interactions and consequences of trade-offs and compromises.

Consensus amongst such diversity will be assisted by knowledge and tools, but long-term solutions will require a shared vision for the tropical savannas.

The components of such a vision enunciated by Stafford Smith and Foran (1993) for grazing management of the rangelands provide a useful guide:

- develop a long-term guiding philosophy for sustainable use of the rangelands;
- place emphasis on quality of production rather than quantity;
- emphasise long-term returns, not annual profit and include an account of land condition;
- plan for cycles in productivity and markets;
- emphasise natural advantages and enhance these with innovation;
- minimise the chance of error, particularly of long-term resource depletion.

Choosing between the almost limitless options for R&D is never easy within an organisation, let alone across the large number of organisations operating in northern Australia. The scientific issues which underpin sustainable use of the soil and land resources have been set down in detail by Williams *et al.* (1993).

Holmes (1993) listed a number of useful questions which need to be addressed for an informed and purposeful R&D program. These include: what uses; by whom; serving which interest group; with what measurable environmental, social and economic outcomes; how compatible with other uses and values; towards what stated national goals; consistency with local, regional, state, national and international obligations?

While there will always be many different contexts against which to frame the scientific questions, progress will always depend ultimately on the quality of our understanding of the key ecological processes operating in tropical savannas. The nutrient, water and energy flows within the ecosystem will need to be characterised and quantified in ways that allow prediction, through time and space, of the consequences of disturbance and intervention on both production and environmental quality. Scale is also of vital importance in tropical savannas. This scientific information will need to be obtained at point, catena and landscape scales and techniques will need to be developed which allow analysis at, and movement between, these scales. While the impacts are often observed at catchment level, the management activity takes place at the level of enterprise.

To be more specific, the Australian Science, Technology and Engineering Council (ASTEC) (1993a) recommended to the Prime Minister's Science and Engineering Council that the following research was required to achieve ecologically sustainable development in the tropical savannas. We believe that such R&D meets the criteria established by Holmes (1993).

Savanna ecology
- the role of woody components in nutrient and hydrologic cycles
- the dynamics of tree/grass interactions,
- catchment level processes,
- determinants of biodiversity,
- indicators of condition and degradation risk,
- the role of variability in landscape resilience and stability,
- the impact of climate change and carbon dioxide levels.

Land use impacts
- the cumulative, long-term effects of grazing,
- the impact of traditional Aboriginal management practices,
- off-site effects of mining and tourism development,
- economics of different land uses.

Integrated management
- methods for integrating information for policy makers to evaluate options,
- predictive models for risk analysis in land use planning,
- strategies for multiple and sustainable land use,
- land tenure to accommodate different economic, environmental, cultural values.

This list should be expanded to include studies within the social sciences, including community education and development and aspects of the sociology of the savanna population.

Among other things, application of this information will:
- give greater meaning and value to baseline data on climates, soils, water, plants and animals;
- provide plausible and acceptable explanations of the current condition and trends of various ecosystems;
- enable the prediction of outcomes of particular disturbances or management practices and a framework for risk assessment of proposed uses by any one of the sectors;
- provide methodologies for the restoration of the productive and ecological potential of degraded systems;
- enable the development of indicators of sustainability and/or ecosystem 'health' from various perspectives;
- contribute towards the resolution of disputes arising from conflicting land use proposals, particularly as social and economic values change over time;
- underpin the definition of degradation which is acceptable and applicable to all the potential clients;
- provide a sound basis for the development of resource management legislation and accompanying regulatory and monitoring protocols;
- contribute toward better targeting of rural adjustment funds to support enterprises which have the greatest chance of both ecological and economic sustainability;
- help Australia meet its international environmental obligations and potentially contribute to improved trade opportunities with 'clean/green' products.

4. MANAGING RESEARCH AND DEVELOPMENT IN THE TROPICAL SAVANNAS

A survey of the extent and nature of research conducted for and in tropical Australia was documented by the NT University for ASTEC (1993b). Research relevant to the savannas is being conducted by some 1200 research units from over 100 organisations comprising State and Territory government departments and agencies, CSIRO, universities and the private sector; about two-thirds of these organisations are headquartered outside the tropics. It is estimated that over 2000 researchers conduct work relevant to northern Australia, spending over $200M annually. The balance of socio-economic objectives of the R&D effort is heavily weighted towards the natural ecosystem and its exploitation by agriculture. Recent growth in R&D providers has come mainly from the university sector and there has been an increasing effort in the social sciences and humanities.

Unfortunately, the research effort is fragmented and often reactive, and usually with a strong discipline rather than system focus. Without change in the nature and coordination of research this effort will not be adequate to achieve ecologically sustainable development. There is no indication at present that any single organisation could mount the multi-disciplinary and long-term research effort required to provide the desired outcome. Even if one such organisation did exist, it is unlikely that any one agency located at a number of sites, could adequately address the complexity of ecological issues within this large and diverse ecozone, let alone the interactions with the potential users and the need for input from economics and the social sciences. Maximal utility of information from this rich matrix of R&D effort will only be obtained from greater interaction, some rationalisation and planned coordination. Additionally, such collaboration is likely to be the only way of accomplishing the large-scale, long-term and multi-disciplinary research identified above.

Substantial benefits could result from greater collaboration and development of better understanding between: (i) the different philosophical attitudes of groups involved with resource management research;

(ii) the savanna user groups; and (iii) the R&D providers and the user groups. Presently, in many cases there is close one-on-one alignment between a particular user group and an R&D provider which may be associated with funding support. While such associations may benefit both parties in the short term, long-term implications for sustainable development from a multi-user perspective may be compromised. Undoubtedly, cross fertilisation of ideas and philosophies will lead to higher quality and relevant research to the benefit of the tropical savannas. The mechanism to provide the much needed rationalisation and coordination could take many forms. What will be essential is for key institutions to come together and develop a collective strategic plan for their research and development in savanna ecosystems. A mix of clients, research agencies and funding bodies involved in developing a strategic plan and agreement for implementation has many of the ingredients of success. A number of these clusters linked into a network may well be the best way forward. The ASTEC (1993a) report developed this concept further and recommended that *'Environmental management models like the Great Barrier Reef Marine Park Authority (GBRMPA) that involve a management agency, regional research institutions, and relevant industry groups should be implemented for other equally important tropical ecosystems'*. The report afforded highest priority to tropical woodland and savanna systems.

The GBRMPA model has many valuable features, particularly the ongoing involvement of stakeholders and the regional basis of the management agency and research groups. Additional or more complex issues will need to be considered if and when such an authority is established for the tropical savannas and woodlands. These include overcoming the current isolation and fragmentation of current groups and the associated parochialism; meeting the different communication needs of the stakeholders; the actual physical occupation of land by competing stakeholders who, for some time to come, will find difficulty recognising the legitimacy of each other's rights; three State or Territory Governments which guard their rights to land jealously and the Commonwealth Government with limited authority to act, compared with one State Government with on-shore authority working with the Commonwealth Government which has off-shore responsibility. Notwithstanding, the sooner a coordinating group with a capacity to influence (i.e. money) is established, the better. It will take a great deal of time and patience to bring a shared vision and plan together. For example, GBRMPA was established in 1975 and it took until 1994 until the 25 year Strategic Plan was released.

6. WHEN SCIENCE CANNOT HELP

No discussion of the role of science in addressing the issue of managing the resources of the tropical savannas would be complete without consideration of what factors other than science influence decision making and development. While this discussion is not comprehensive, we consider that it will be the social, political and economic issues which will have the greatest effect on the use of tropical savannas in the short-term. The scientific principles which underpin ecological sustainability may be neglected in the short-term, but in the longer term it will be these biophysical laws that will determine the condition of the savanna for the land use policy adopted by the society.

a. Politics

Australia is not alone in experiencing Government initiatives made for political or other expedient reasons which do not have a sound ecological or economic base. Dams and agricultural schemes are the most obvious examples and there are other examples in the mining, tourism and social sectors. Examples in the agricultural, forestry and fishing sectors have been best articulated by Blaikie (1985) in his book entitled *The political economy of soil erosion in developing countries*. These include the use of funds to promote short-term financial gain to a particular sector or region rather than longer-term social and/or environmental benefits to the wider community; targeting funds to capital works which have a high visibility and are achievable within political timeframes of 3-5 years; failure to recognise that needs are often greatest for groups of people or regions with low political influence; over-attention in policy development to the needs of the economically or socially powerful groups, not always to the best effect on resource use.

b. Sectoral interests

Over time, different sectoral groups capture the public agenda. In northern Australia the pastoral sector has enjoyed widespread community support for many years, epitomised by the thrust to tame and develop the country, as exampled by the brigalow scheme in southeast Queensland. Later, this sector was joined by the mining industry as Australia expanded its resource-based economy. Nowadays the 'green' sector has been legitimised and plans for clearing trees, introducing new plants for increasing agricultural production, and development of mines in environmentally sensitive areas are thoroughly examined before approval. Recent initiatives by the Federal Government have significantly elevated the sectoral interests of Aborigines to a level not considered possible only 5 years ago.

c. Rights

The issue of sectoral interests naturally leads to that of attitudes and rights. The debate is currently in a state of flux, where separate groups are reluctant to accept the legitimacy of the attitudes and rights of others. The graziers espouse their rights perforce occupancy, leasehold agreements and the 'right to earn an honest living', certain public groups herald their rights as the custodians of the ecosystem for future generations,

Aboriginals proclaim historical and cultural rights, the mining sector declare their rights to generate wealth for shareholders and the nation, whilst all citizens of the nation actually possess the rights, expressed through State and Commonwealth Governments. Governments will pay greater or less attention to the rights of these sectors for many other reasons than ecological sustainable development! The 'rights' issue has been further complicated now that these Governments have agreed to meet international obligations which, in effect, sign away some of our national rights to the rest of the world.

d. Knowledge

The scientific process will always lead to established theories and ideas being replaced as new data and analysis confirm new understandings. Unfortunately this process is vulnerable to ridicule and the discrediting of scientific evidence in the debate over conflicts in resource use. In such circumstances scientists can, of course, appear to be their own worst enemies. They argue and disagree about aspects of their work and frequently publish information which refutes previously held theory. This apparent confusion is often the basis for discrediting scientific evidence which is unacceptable to one or other sector, particularly those outside the scientific community. Commonly held beliefs in such things as climatic cycles and 'natural' biophysical processes are often maintained in the face of high quality scientific evidence to the contrary when expedient to a particular cause.

e. Refraction

Last, but not least, is the issue of data refraction, that is, when the same information can be seen to support opposing viewpoints. Two examples for the tropical savannas come to mind. Under certain grazing pressure individual animal weight gains from degraded pastures can be higher than from non-degraded pastures. The grazier recognises higher economic returns from the degraded pastures, while the ecologist is aware of the ecological damage and risk of greater degradation if grazing pressure is not very carefully managed. A second example is the increase of woody weeds in many of the tropical savanna pastures. In some situations pastoralists benefit as animal production may be increased by the presence of these plants as they can provide a good 'top-feed' reserve of dry season feed and/ or a source of scarce protein and shade. However, management of woody plants in such communities is very difficult and has rarely been achieved. Scientists recognise that the most likely outcome is reduced production and an ecologically degraded shrubland.

CONCLUSIONS

The time is ripe for the diverse group of land users of tropical savannas to come together to develop a joint vision for sustainable development of these ecosystems over the next 20–50 years. These groups could generate a common set of management principles and guidelines, based on current knowledge, which identify how the vision might be achieved. Such a coming together could also be used to develop a coordinated strategic plan for research, development and implementation of long-term sustainable land use of the savanna ecosystems. This interaction will identify areas of common interest and concern and will enable a greater appreciation of the different perspectives and viewpoints of savanna users. This will not be a simple process and the temptation for political point-scoring must be avoided.

The development of policies and strategies cannot wait for all the information to be gathered. Innovation must be used where there is uncertainty! The formation of partnerships between the land managers, the government regulatory agencies and the research and development agencies will be essential. The two-way flow of information and innovation between the research and development agencies and those directly responsible for land management must be greatly increased.

The R&D sector must learn to work more co-operatively and collaboratively. This change must not only apply within disciplines but more widely across disciplines, since achievement of many of the planned management and ecological outcomes will depend upon the application of information generated by social scientists and economists.

Finally, tropical savanna land users need to be proactive and bold, rather than reactive and passive. Tropical savanna users have the unique opportunity to set the agenda together, rather than responding to an agenda set by others which might disenfranchise or alienate one or more sectors.

The challenge is before us. Maybe this symposium will be one of the first steps upon the long journey to sustainable use of our tropical savannas.

References

Ash, A.J., McIvor, J.G., Corfield, J.P. and Winter, W.H. (1996). How land condition alters plant-animal relationships in Australia's tropical savannas. *Agriculture, Ecosystems and Environment* (in press).

ASTEC (Australian Science and Technology Council) (1993a). *Research and Technology in Tropical Australia and their Application to the Development of the Region.* AGPS, Canberra.

ASTEC (Australian Science and Technology Council) (1993b). *Research and Technology in Tropical Australia: Survey.* Occasional Paper No.24. ASTEC and Office of Northern Development, Canberra.

Australian and New Zealand Environment and Conservation Council Task Force on Biological Diversity (1993). *National Strategy for the Conservation of Australia's Biological Diversity.* AGPS, Canberra.

Blaikie, P. (1985). *The Political Economy of Soil Erosion in Developing Countries.* Longman Scientific and Technical, Essex.

Ecologically Sustainable Development Steering Committee (1992). *National Strategy for Ecologically Sustainable Development*. AGPS, Canberra.

Holmes, J.H. (1993). The changing context of rangeland research and development: a socio-economic perspective. In: *R&D for Sustainable Use and Management of Australia's Rangelands*, (eds Morton, S.R. and Price, P.C.), pp.34-39. Land and Water Resources Research and Development Corporation, Canberra.

Hutchinson, K. (1992). Evaluating long term sustainability in rangeland environments. In: *Environmental Indicators for Sustainable Agriculture*, (ed Hamblin, Ann), pp37-44. Bureau of Rural Resources, Parkes, ACT.

Ive, J.R. and Cocks, K.D. (1988). LUPIS: A decision-support system for planners and managers. In: *Desktop Planning: Advanced Microcomputer Applications for Physical and Social Infrastructure Planning*, (eds Newton, P.W., Taylor, M.A.P. and Sharpe, R.). Hargreen Publishers, Melbourne.

MacLeod, N.D. and Taylor, J.A. (1994). Community perceptions of grazing impacts in the beef producing regions of Queensland. *The Rangeland Journal*, 16: 238-253.

McIvor, J.G., Ash, A.J. and Cook, G.D. (1995). Land condition in the tropical tallgrass pasture lands. 1. Effects on herbage production. *The Rangeland Journal*, 17: 69-85.

Mentis, M.T. and Seijis, N. (1993). Rangelands bio-economics in revolutionary South Africa. In: *The World's Savannas*, (eds Young, M.D. and Solbrig, O.T.), pp.179-203. UNESCO, Paris and Parthenon Publishing Group, Carnforth.

Stafford Smith, D.M. and Foran, B.D. (1993). Problems and opportunities for commercial animal production in the arid and semi-arid rangelands. In: *Proceedings of the XVII International Grasslands Congress*, pp.41-48. New Zealand Grassland Association, Palmerston North, NZ.

Standing Committee on Agriculture and Resource Management (1993). *Sustainable Agriculture: Tracking the Indicators for Australia and New Zealand*. Report No.51. Agricultural Council of Australia and New Zealand. CSIRO, Melbourne.

Veitch, S. (1992). Integrated indicators and modelling scenarios: an example with the 'WhatIf' computer package. In: *Environmental Indicators for Sustainable Agriculture*, (ed Hamblin, Ann), pp54-59. Bureau of Rural Resources, Parkes, ACT.

Williams, J., Helyar, K.R., Greene, R.S.B. and Hook, R.A. (1993). Soil characteristics and processes critical to the sustainable use of grasslands in arid, semi-arid and seasonally dry environments. In: *Proceedings of the XVII International Grasslands Congress*, pp.1335-1350. New Zealand Grassland Association, Palmerston North, New Zealand.

Chapter 3

Changing Resource Values in Australia's Tropical Savannas: Priorities in Institutional Reform

John H. Holmes

*Department of Geographical Sciences & Planning,
The University of Queensland,
St. Lucia, QLD., 4072
Australia*

Abstract

- More so than in any other Australian bioclimatic zone, rapid changes in both market and non-market resource values are propelling a major shift in land ownership and land tenure institutions.

- Shifts in resource values do not translate into income streams for private landholders: in effect, a widening gap between value and private title. This divorce is leading to 'deprivatisation' of land ownership with over one-fifth of pastoral lands in the northern zone already transferred to other titles or to non-private ownership.

- Changing resource values are leading to a sharper regional differentiation between: core pastoral regions with only modest changes in resource use and land tenure; marginal regions experiencing rapid change in values, uses and tenures; and urban-oriented development regions.

- Institutional change has been occurring in a piecemeal manner, with inadequate attention to underlying causes, current trends and long-term outcomes. A more systematic, informed approach is needed to ensure an effective match of property rights (and use rights) with resource use, including: clearer specification of the rights of both titleholders and non-titleholders whether on Aboriginal or on other tenures; more attention to mechanisms for assisting land use change; and possible creation of a new form of public tenure capable of accommodating private use rights.

- In keeping with an enhanced public role in resource allocation and conflict avoidance, there is a growing need for strategic regional planning, focusing mainly on marginal regions and urban development regions. Strategic planning needs to be linked to changing land tenures and uses as well as to more informed and sensitive approaches to environmental, economic and social impact assessment.

- However, successful regional planning will only emerge following an extended learning period, given the novelty of the context, the rapidity and unpredictability of change and the mix of interests meriting a role in the planning process.

Table 1: Northern Tropical Savanna Zone: Change in area held in privately-owned pastoral leases: 1974–1994

	Area in Privately Owned Pastoral Leases (km²)			
	Western Australia	Northern Territory	Queensland	TOTAL
1974	82 112	218 653	285 880	586 645
1994	46 552	157 556	250 985	455 103
Change	−35 560	−61 087	−34 895	−131 542
Percent Change	−43.3	−27.9	−12.2	−22.4

1. INTRODUCTION

Earlier, I argued that we need to undertake a far-reaching reconceptualisation of resource values in Australia's tropical zones (Holmes 1995, see Chapter 1). We need to recognise a shift away from commodity-oriented values towards people-oriented and environmental values, with the shift arising not only from past disappointments in resource development, but being driven by changing national perspectives and aspirations. These, in turn, are indicative of a wider global trend, loosely described as the transition from an industrial to a post-industrial society. Parallel shifts are occurring in comparable, marginal, frontier zones in other western nations. This shift is accompanied by a rapidly changing balance of political and economic power among interest-groups, both within and beyond the tropical savannas, with a growing diversity of interests claiming a role in determining resource issues.

In this paper, I argue that changing values are requiring radical changes in resource-related institutions. Most striking have been changes in land ownership and in land tenures through recognition of Aboriginal land rights, but there are other changes underway, including the retreat of pastoral tenures from the most marginal lands, and the parallel search for new tenures to accommodate non-pastoral uses. There has also been gradual recognition that governments have newly emerging responsibilities influencing future land use, with tentative moves towards strategic regional planning.

Governmental responses to these new challenges have been fragmentary and belated, with an over-reliance on legalistic 'solutions', most notably in the award of Aboriginal land rights. There is scope for more comprehensive, coordinated approaches, to ensure that present challenges are seen as opportunities for ensuring effective institutional restructuring to meet future needs.

1.1 Institutional responses: causes and outcomes

It is already clear that a major restructuring of land tenures and property rights is occurring in response to new resource values, involving a reversal of the historical trend towards enhanced private ownership of land-based resources, and an emerging, powerful thrust towards 'deprivatisation'. This thrust is having equally powerful impacts on such pivotal institutions as land tenures, land ownership and property rights, which are discussed in detail in this paper. We need to recognise the overall dimensions of this trend and to understand the structural economic changes which are impelling this trend.

As shown in Table 1, in the northern savanna zone, over 22% of privately-held leasehold land has been transferred into other forms of ownership over the last two decades, with the chief transfers being towards Aboriginal ownership and National Parks, but also with major transfers to various state and federal governmental agencies. This trend, which began in the 1970s, gathered momentum in the 1980s and early 1990s and appears likely to continue for some time. In the northern savanna zone, land under privately held pastoral lease has diminished to encompass less than 54 percent of the zone, and may well decline to be less than fifty percent. The current distribution of land tenures is shown in Figure 1, with the areas held in each category shown in Table 2.

This deprivatisation thrust is most evident on marginal lands, now incapable of generating adequate income-streams to private landowners, but it may also extend to encompass some lands of continuing, if modest, pastoral value.

This process of deprivatisation is a logical response to changing resource values, consistent with current property-rights theory, particularly as proposed by the pragmatic, institutional school of resource economists (Bromley 1991). Just as it is sensible to privatise resources where they can be efficiently utilised in private ownership and are capable of generating income-streams for owners, so also it is appropriate to deprivatise where income streams are no longer readily tied to private title, and where there are increasingly dysfunctional economic and social outcomes from private ownership. The negative outcomes have been discussed in Holmes (1991), while the growing divorce of resource values from private title, including both market and non-market values, has been discussed in a

Table 2: Land Tenures in the Northern Savanna Zone: April, 1994

Tenure	Western Australia (km²)	Northern Territory (km²)	Queensland (km²)	TOTAL (km²)	(%)
Aboriginal Freehold	-	154 150	18 670	172 820	20.4
Aboriginal Reserve	26 280	-	11 370	37 650	4.5
Aboriginal-owned pastoral lease	20 006	15 306	9868	45 180	5.3
Private pastoral lease	46 552	157 566	250 985	455 103	53.8
Freehold	-	6400	1600	8000	0.9
Vacant Crown Land	27580	500	200	28 280	3.3
Government-held leases	4044	16 193	1840	22 077	2.6
Other Crown land	850*	900*	8200	9950	1.2
Defence land	5660	3285	-	8945	1.1
Mining Reserve	-	-	3400	3400	0.4
National Park	10 220	24 900	20 067	55 187	6.5
Total	141 192	379 200	326 200	846 592	100

recent paper (Holmes 1994a) and needs only brief summary here.

1.2 Increasing significance of non-market values

The most readily discernible trend is the increasing significance of non-market values, listed in Table 3. Of these, the one which is having the most substantial, widespread impact is the recognition of Aboriginal land rights, with a strong, continuing momentum to transfer pastoral land into Aboriginal ownership. Increasingly, Aboriginal lands are being passed into distinctive nontransferable titles, whether it be Aboriginal freehold, as in the Northern Territory and Queensland (also in South Australia, outside of the tropical savannas) or the newly-legislated native title, which will be available for eligible claimants where the land has been purchased on their behalf.

Transfer to Aboriginal ownership does not necessarily spell the end of pastoralism, nor the demise of market-oriented, production. Indeed, this is highly unlikely on

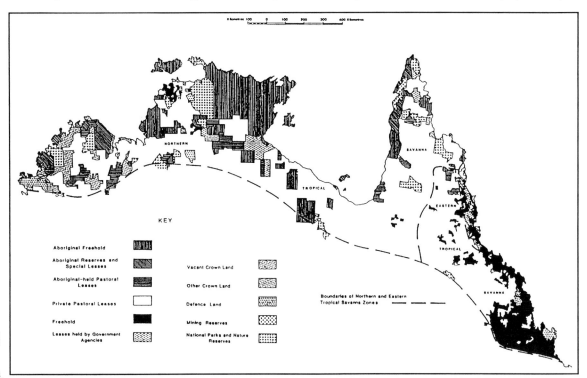

Figure 1. Land tenure in Australia's tropical savannas: 1994.

Table 3: Rangeland Non-Market Resource Values

• USE VALUES
Aboriginal uses
Conservation
Non-commercial recreation
• EXISTENCE VALUES
Biological/Ecological Values
Landscape Values
Aboriginal cultural Values
Non-Aboriginal cultural Values (including pastoralism)
• OPTION VALUES
Diverse array of identified and yet-to-be identified values

Table 4: Rangeland Market Resource Values

• TIED TO BROADACRES LAND TITLE
Pastoralism
Limited tourism and recreation
• TIED TO SMALL LAND PARCELS
Resort-based tourism
• TIED TO NON-LOCAL ENTERPRISES
Various footloose activities (see below)
• FOOTLOOSE
Mobile resources: Hunting
Mobile enterprises: safari tourism
Footloose consumers: private tourism

acquired lands with adequate pastoral potential. However, transfer to Aboriginal ownership has two critical structural outcomes: firstly, commercial pastoralism will be subordinated to other traditionally-oriented values (particularly, on lands of low pastoral potential, as is usually the case); and, secondly, the nontransferable title, with its group-ownership ensures that the land is not treated as a marketable asset.

Quite apart from Aboriginal land rights, there are other non-market values which are of increasing national significance, with the most notable being conservation goals, linked to the national biodiversity strategy. Bridgewater and Walton (1995) have emphasized that the tropical savannas are a 'complex, heterogeneous landscape', which has been subject to much less modification than most other Australian bioclimatic zones, with significant benefits in adopting a holistic, landscape-oriented approach, involving compatible management of lands outside native reserves. This adds considerably to the non-market values attached to land currently held in private tenures.

Less readily apparent, but possibly of comparable long-term significance, are the structural changes in market values, which are leading to a growing divorce between these values and broad-acres private title. In Table 4, current market values for the rangelands are listed, according to their degree of attachment to (or detachment from) broad-acres land title. The only resource value with strong ties to land title is pastoralism. This is not the case with any of the newly emerging market values. Even with tourism, on which some hopes are raised as a complementary activity to pastoralism, economic benefits are not closely tied to broad-acres land title. While many outback tourist and recreation activities are highly reliant upon broad-acres resources, these activities do not readily generate income to the landholder, either because they are footloose or else linked to highly localised capital investment in resorts and facilities which have very limited land demands. Major natural attractions such as beaches, estuaries, rivers, wetlands, gorges, ranges and natural and cultural heritage sites almost invariably remain in public ownership, for reasons not only of cultural tradition but also economic benefit. On efficiency grounds, as well as equity, these resources can appropriately be regarded as public goods. Indeed, the greatest opportunities for expanded economic activity in private-sector tourist-related activities are often strongly tied to public ownership and/or management of natural attractions for tourism and recreation, such as Kakadu, Nitmiluk, Lawn Hill and Lakefield.

Thus, in a seemingly perverse way, the 'invisible land' of the market, so frequently invoked by free-market theorists, is working towards deprivatisation of much of the tropical savanna lands. Because of their incapacity to generate income-streams to landholders, the marginal lands are losing market-value, which facilitates their transfer to other tenures, and, in some cases, is obliging state and territory governments to acquire land title virtually by default or as a means of averting undesired outcomes.

While market forces and political decisions seem to be working in concert to achieve needed institutional change, nevertheless the process is imperfect. Inevitably any context involving rapid changes in values and in institutions will lend itself to distortions and to opportunistic behaviour which impedes change. Currently, in north Australia, needed changes are being impeded by: continuation of traditional optimism about opportunities for development of agricultural and pastoral commodities; imperfect perception of basic processes of structural change and lack of awareness of their implications; relinquishment by governments of their traditional constraints on property rights attached to pastoral leases; and a consequent niche for rent-seeking activity by opportunistic speculators, concerned with wealth-capture rather than wealth-creation, from the process of change in resource use and land tenure.

2. REGIONAL IMPLICATIONS OF CHANGING RESOURCE VALUES

Regional differentials in income-earning capacity from broad-acres land use are still tied to pastoral productivity, with the most critical consideration being capa-

bility to support a managed cattle herd. This need was highlighted during the Brucellosis and Tuberculosis Eradication Campaign (BTEC) and has been reinforced by subsequent trends in livestock production systems.

Recognising the inability of marginal pastoral lessees to meet BTEC standards, the Northern Territory Department of Lands and Housing commissioned me to undertake an assessment of the economic viability of thirty pastoral leases in the Gulf District. This involved a property-by-property assessment of land attributes which influenced income generation and capacity to manage the herd. Of these attributes, the most important was carrying capacity, with three beasts per square kilometre being assessed as the minimum value in order to meet basic capital and operating costs. Also taken into consideration was the degree of difficulty in mustering and in constructing and maintaining fences, together with within-property accessibility. On the Gulf Plain only scattered pockets of land had the capability to support a managed herd, and these were generally too small and isolated to ensure economic viability (Holmes 1990).

The national map of livestock densities suggests that large tracts of the northern tropical savannas do not have the needed levels of cattle densities to support effective herd management. This is indicated in Figure 2 of Holmes (1995, see Chapter 1, this volume). It must be noted that there have been subsequent major herd reductions, mainly, but not entirely, as a result of BTEC.

Herd manageability and economic viability are further impeded by cost burdens arising from remoteness and lack of infrastructure. Laut and Nanninga (1985) give a regional appraisal of environmental and locational variables influencing the economies of cattle production in the tropical savannas.

Pastoral lease rentals can be regarded as a crude composite index of pastoral values and production costs. The map of 1990 lease rentals for Cape York Peninsula is shown in Figure 2. While the map does show considerable local variability, this is overshadowed by the very sharp rental gradient by latitude, with southern leases commonly having rentals per square kilometre ten times higher than northernmost leases. For leases exceeding 1000 km², there is a forty fold variation, ranging from $2.40 (Wrotham Park) to $0.06 per square kilometre (Shelburne Bay). The recent conversion to unimproved capital valuations has not significantly altered this rental gradient, save only for a modest enhancement of the lowest rentals, in recognition of a continuing speculative element in their market values.

These regional contrasts must be fully recognised, if only to ensure that institutional restructuring, now underway, is accomplished in a manner reasonably compatible with the geography of resource values in the tropical savannas. A useful starting point is to recognise three distinct regional types, each requiring a distinctive approach towards institutional reform. These are:

1. *Core pastoral regions*, where cattle grazing will continue as the dominant broad-acres land use. These are the regions where productivity has been sufficient to sustain needed investments to ensure effective herd management, reflected in a positive response to BTEC investment subsidies. These lands coincide with the Tropical Tallgrass Savanna, encompassing the Fitzroy, Burdekin and Einasleigh regions and adjacent east coast lands, where higher productivity levels are widely acknowledged (Mott et al. 1985). Other regions where substantial tracts of land can sustain the requisite productivity levels include southern Cape York Peninsula, Queensland's Gulf District, the Barkly Tableland, Victoria River District and Southern Kimberley, with the most productive areas generally located to the inland of the tropical savanna zone.

2. *Marginal, or 'frontier' regions with diverse resource values*: These encompass most of the Monsoon Tallgrass Savannas, described by Mott et al. (1985), including areas considered unsuited to pastoralism, and until recently held mainly as Aboriginal reserves or vacant crown land. Also included are marginal pastoral leases found to be incapable of satisfying BTEC standards, and with their historically small scattered cattle numbers now severely depleted. While these regions offer a tantalising array of resource values, few seem capable of generating income for local peoples, many of whom rely directly or indirectly on government transfer payments. Income-generating resource values are highly localised, arising mainly from mining, agriculture and tourism. These regions include northern Cape York Peninsula, the Northern Territory Gulf District, Arnhem Land, the outer Darwin hinterland and the north Kimberley.

3. *Urban-oriented development regions*: These are regions where land-related resource values are being transformed by enhanced accessibility and by population numbers, generating demands for land which require distinctly different policy responses and institutional structures compared with the frontier regions. Eastern Queensland's coastal savannas have for long been valued more highly, not merely for their inherent productivity potential but also their high accessibility and proximity to major urban centres, such as Townsville. Darwin's immediate hinterland is now experiencing a comparable revaluation of land resources, creating a strong market for various classes of land, particularly small land parcels for residential, recreational and other purposes.

The Future of Tropical Savannas

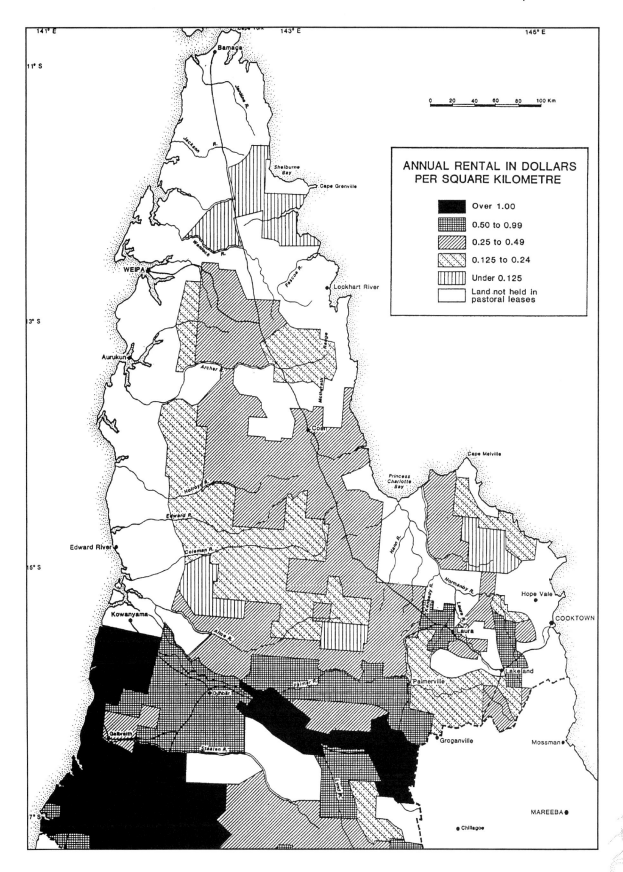

Figure 2. Annual land rental in the Cape York region of Queensland

The Northern Territory Department of Lands and Housing has engaged in comprehensive strategic planning in response to these demands. Pressures may emerge around other urban nodes, with the most probable growth points being Katherine and Kunnunurra, together with further expansion of the regional land market centred in Cairns, extending into savanna lands north of Cooktown (a prospect which was early recognised by the current owner of the Starcke leasehold and freeholded lands).

3. Needed institutional reforms

Existing institutions for the development and delivery of public policies and programmes were designed for an earlier era and are ill-suited to emerging needs. As already stated, the two most urgent areas for reform are concerned with property rights and with strategic regional planning. In both these areas there have already been significant, even radical, innovations, but these must be regarded only as forerunners to further needed change.

So far, the most radical changes have been in the area of property rights, most notably in the legislative and judicial recognition of Aboriginal land rights. In the legislative arena, the landmark *Aboriginal Land Rights (N.T.) Act, 1976*, represented a historic shift in two important respects. First, it gave legislated recognition to Aboriginal land-ownership and created an entirely new form of land title (or property right) involving group, nontransferable ownership of both surface and subsurface resources. Second, it enabled Aboriginals to acquire ownership of more than forty percent of the Northern Territory, admittedly almost entirely comprised of unwanted, vacant Crown land or former Aboriginal reserves. In the judicial arena, the High Court's Mabo decision, 1992, resurrecting native title is universally acknowledged as a radical reform, again concerning property rights. These two decisions, together with independent or parallel legislation within states and the Northern Territory is leading to dramatic changes in the map of Australian land tenure, primarily in the marginal lands of the arid interior and the northern tropical savannas. The regional concentration of land tenure changes is well revealed in the 1993 Australia Land Tenure Map (AUSLIG 1993), with contrasting regional outcomes between the northern and northeastern tropical savannas being shown in Figure 1.

There are other fragmentary moves towards institutional reform in recognition of changing priorities on the ownership and use of resources. Examples include: Aboriginal ownership with joint management of national parks; expansion of national parks and reserves; reform of lease tenures, replacing development and stocking conditions with sustainable use requirements; examination of possible alternative tenures for non-viable pastoral holdings; new mechanisms to facilitate appropriate non-pastoral uses. Overall, there is a persistent trend towards recognising the retreat of pastoralism from the least productive lands and selectively recognising and encouraging a variety of non-pastoral uses, through a reshaping of property rights, as specified in lease titles and permits.

The second major thrust, is towards regional strategic land use and development planning. Such activity would have been redundant in an earlier period when pastoralism was granted unfettered priority in broadacres land use. The move towards comprehensive planning is a response to the challenges posed by multiple resource values, diversification in interest groups and policy options, uncertainties concerning land use and regional development directions and the continuing pivotal role of governments. These are characteristics currently encountered in marginal, 'frontier' regions in western nations, with the most striking parallels being found in northern Canada and Alaska (Holmes 1992).

In rapid succession, three of Australia's most remote and least developed northern regions have become the focus of strategic regional planning. The *Kimberley Region Plan Study Report*, prepared jointly by the Western Australia Department of Regional Development and the Northwest and Department of Planning and Urban Development, was released for public comment in December 1990. The *Gulf Region Land Use and Development Study* was published by the Northern Territory Department of Lands and Housing in October 1991. Finally the most ambitious proposal is the Cape York Peninsula Land Use Strategy (CYPLUS) initiated by Queensland government in early 1990, and with joint state-federal involvement, comprising a three-stage programme extending over a five-year period.

In the remainder of this paper, I will focus on the two central issues of property rights and regional planning. Property rights issues have become increasingly significant and controversial in contemporary society, particularly so in rangelands and other marginal zones. A review of current debates in the American, New Zealand and Australian rangelands can be found in Holmes (1994b). Also, as pointed out already, regional planning is becoming increasingly important in these zones, and a discussion on alternative approaches to the planning task provides an effective framework for discussing related public institutions, decision-making processes and modes of implementation.

Since it is the marginal, or frontier, regions where the need for institutional reform is most evident, my discussion will focus mainly on these regions. However, it would be unwise to engage in a strict separation of the marginal from the core pastoral regions, particularly on regional planning issues. In some cases, appropriate geographic entities for planning purposes must include both categories. This is clearly so with Cape York Peninsula and the Kimberley Regions, where there is a high level of linkage, or functional integration between core

pastoral and marginal lands. Appropriately, the regional planning exercises undertaken in these two cases have included both core and marginal lands. On the other hand, the Northern Territory Gulf and Barkly Regions can readily be separated for planning purposes, simply because functional ties are tenuous.

4. Land tenure reform: restructuring property rights

For the tropical savannas, there are two major arenas involving land tenure reform. The first of these concerns Aboriginal land rights, now only at mid-stage in the reshaping of land tenures and the reallocation of lands. The second arena concerns the property rights and duties attached to pastoral lease tenures, and any successor tenures now needed to accommodate a diversity of non-pastoral values.

1. Aboriginal land rights

Aboriginal land rights issues are far too wide-ranging and complex to receive proper treatment in this paper, and I mention only two issues of particular relevance to the overall management and use of tropical savanna lands.

The first concerns the determination of lands to be transferred to Aboriginal title. While circumstances hardly permit otherwise, given the outcomes from the history of land occupation in Australia, the award of Aboriginal land rights is being decided almost entirely by legal determinations, leading to inequitable outcomes between Aboriginal groups.

Unlike north Canada, where there are no pre-existing non-native property rights to restrict current comprehensive land claim settlements, in north Australia the prior award of private titles, mainly pastoral leases, has acted as a constraint on land transfers. Some measure of equity may be achieved through selective purchase of pastoral leases now likely to be accelerated with the establishment of the Aboriginal and Torres Strait Islander Land Fund legislation, introduced into federal parliament in June 1994, and by the prospective transfer of this land to native title under the provisions of the federal *Native Title Act, 1993*. However, any such programme is likely to be pursued in isolation from wider regional land use objectives, while also raising controversial issues about the appropriate land title for the purchased lease.

A second matter for concern is the separatism and social division promoted by current land legislation. The Aboriginal freehold title being issued is simultaneously more powerful and more restrictive than those available to non-Aboriginal people who also live in the region. This reinforces the dualism between Aboriginal and non-Aboriginal lands and the associated social divisions are strongly felt locally; and it also strengthens the growing resistance towards recognition of further Aboriginal land claims, such that the lack of any compromise land title may exclude some Aborig-

inal groups from acquiring any land. Accordingly, it may reinforce and perpetuate inequitable outcomes for Aborigines, and preclude multiple land-use options that require shared decision-making between Aboriginal and non-Aboriginal representatives.

Apart from joint arrangements for the management of a few national parks, little thought has been given to the possibility of incorporating Aboriginal interests within a more comprehensive review of land tenure and land use directions in northern marginal regions. While any change of strategy, at this stage, may well raise more problems than it solves, nevertheless it is sensible to examine options, some of which may achieve more useful long-term outcomes, particularly on marginal lands. The desired outcome is to expand Aboriginal rights to regional resources, beyond those within the boundaries of Aboriginal freehold lands, preferably with minimum impairment of other interests. Given the resistance to further expansion of Aboriginal freehold, there is a strong case for considering some intermediate forms of land tenure, enabling Aboriginals to acquire rights to a more limited set of resources, but over much larger areas than currently seems politically feasible. This intermediate form of tenure would allow more flexibility in land use planning, would expand the Aboriginal resource base without major withdrawals from non-Aboriginal interests and would assist towards reducing community separatism. These rights could be incorporated selectively into new forms of tenure described below. While primarily directed towards access rights to engage in traditional activities, as already incorporated into legislation on pastoral lease tenures in Western Australia, Northern Territory and South Australia, these proposed tenures could provide additional rights, but falling short of those currently awarded under Northern Territory Aboriginal freehold title. It is assumed that the rights awarded to the Aboriginal groups would be derived from their traditional association with the land in question and would not be marketable.

2. Lease tenure reform on core pastoral lands

Since I have recently written extensively on rangeland tenures (Holmes 1991, 1994a, 1994b, Holmes and Knight 1994) all that is needed here is a summary of the main ideas, also focusing on their relevance to the tropical savannas. Of the ideas presented in these papers, the most important is to recognise mounting evidence on the need for entirely different land tenures between the core and the marginal grazing lands. The case for differentiated broad-acres tenures is strongest in the tropical savannas, where the resource values of the most marginal lands, whether market or non-market, are not readily attached to land title. This issue is discussed separately in the following section.

For the core grazing lands, the case for retention of lease tenures, with clear specification of the rights and duties of lessees has been pursued both in terms of the pragmatic needs of the various interest-groups and the

general public interest. It is also consistent with current directions in property rights theory (Holmes and Knight 1994). An appropriate lease tenure will allow the titleholder to make effective use of the pastoral resource, while reserving other rights to the state, selectively to be made available according to emerging resource needs. In summary, we have argued that the core attributes of an appropriate pastoral lease tenure system are as follows:

(a) *Clear specification of the property rights of the titleholders*, designed to meet the resource needs of pastoral enterprises;

(b) *Performance standards or duties attached to the titles*, with an increasing emphasis on sustainable land use;

(c) *Capacity to award additional property rights* where additional resources can be internalised effectively, with the additional right being either purchased or rented;

(d) *Specification of the rights of other interest-groups*. These might include access rights for the public to engage in recreational activities, Aboriginal groups to engage in traditional activities, or miners to explore and develop mineral deposits, or preservation covenants/conditions;

(e) *Retained powers of resumption for more intensive, higher-rent private uses* in addition to public purposes, with full compensation for the loss of pastoral values;

(f) *Retained powers to revise lease conditions at regular intervals*, with compensation payment and/or rent reduction for any reduction in pastoral values;

(g) *Payment of an annual rent* in recognition of the tenure status, where the rental ratio is linked to the apportionment of resource asset-values between the lessor and the lessee.

The historical record of Australian land settlement shows that these core attributes were selectively used as powerful policy instruments during the interventionist periods of frontier development and government-sponsored closer settlement, even up to the 1950s, in New South Wales and Queensland, only to fall into disuse as policy momentum was lost. However, the recent revival of public interest in the rangelands is leading to a parallel revival of interest in the capabilities of lease title as a policy instrument, with initial attention focusing on discarding now-obsolete conditions on minimum stocking rates and capital investment and their replacement with range monitoring and range care objectives. South Australia and the Northern Territory have taken the lead in incorporating these new objectives within the tenure system, but the other states all show signs of moving in this direction.

Quite apart from the side issue of rentals (which arouses heated controversy, disproportionate to the modest sums involved) there are three other major policy areas, within which lease tenures can act as effective policy instruments, namely: access; preservation of biodiversity; and conversion of pastoral land to other, more intensive uses. We have argued that the lease tenure mechanism has considerable potential to act as an effective policy instrument in pursuit of emerging policy goals in these three important areas. Currently these issues are being addressed in piecemeal fashion within most jurisdictions, and the outcomes are likely to be less satisfactory than they would be if a comprehensive approach to lease tenure reform were adopted, with the final outcomes being partly based upon negotiated trade-offs between the lessor and the lessees.

Whether reforms are undertaken comprehensively or in piecemeal fashion (as is more likely), there are advantages in including a learning period in which draft policies are tested, preferably adopting a low key approach where action is taken only to resolve conflicts or to recognise significant changes in resource use. I mention here two examples of a pragmatic approach to reform, both from the Northern Territory, with one relating to the issue of public access and the other to procedures for approving non-pastoral uses.

Northern Territory's recent legislative reform of lease tenures provides for public access across leasehold land to any permanent natural water or to any site of recognised public interest. The adopted policy is to allow lessees, if they wish, to designate routes, within a given time period. Thereafter, the responsibility rests with the Pastoral Board, but the Board will act only where actual or potential conflict arises between an aggrieved lessee or an aggrieved third party. This policy is based upon the seemingly well-founded assumption that informal arrangements, locally negotiated, will prove adequate, particularly since lessees now need to take into account the reserve powers held by the Pastoral Board. In due course, the issue of payment for access will have to be considered. A logical extension of the present policy approach would be to accept the right of the lessee to levy a payment, but only in return for the provision of facilities and support services. Again, a learning period, based upon informal arrangements, but with the Board having reserve powers to make determinations, would seem to be appropriate.

The second emerging major policy area is in the inclusion of non-pastoral enterprises on pastoral leases. Again, a pragmatic approach, now being proposed in the Northern Territory, has many commendable features. The Department of Lands and Housing is drafting a comprehensive, sequential system for the granting of additional non-pastoral rights geared to the resource use, level of inputs, level of involvement for parties other than the lessee and degree of complementarity with pastoralism. Protocols are being drafted, specifying non-pastoral uses which may be undertaken by the lessee without requiring approval of the Pastoral Board, also those for which a permit or licence will be required, as well as those requiring a separate lease.

This allows a progressive expansion of property rights commensurate with expansion in resource use, with lapse of permits and, for a specified time limit, of leases, where relevant resource use lapses.

In the award of additional property rights, a central principle is that the beneficiary must pay for these rights, just as the removal of rights must attract compensation. Licences and permits usually require annual fees, while special leases can either be purchased or attract annual rentals. Of course, where only modest income is generated, rents or fees should be modest, but beyond a given threshold value, can be scaled against annual income or unimproved capital values. Give-away prices for the award of additional rights can only generate speculative booms, and even fraudulent deals, as evidenced by the low-cost free-holding of prized land tracts on Cape York Peninsula.

The third major policy area where lease tenures can be effective policy instruments is in the preservation of biodiversity, now formally recognised as a national goal. The importance of the specificity of covenants, attached to individual leases can be recognised when considering the proposals developed in a landmark paper by Morton et al. (1995) who argue for the integration of conservation and production in the 'semi-natural matrix' of arid Australia. They point out that conservation needs can not be met solely by conservation reserves and that off-reserve protection is essential. A 'regional landscape approach' is needed for the maintenance of biodiversity, with special attention given to the management of critical 'resource-rich patches.' The authors propose a hierarchy of tenures in the pastoral region, with excised management units, restricted use units, and sustainable use areas. They also suggest that the titleholders receive financial assistance for the costs of managing conservation values as this would help to reconcile economic pressures with ecological sustainability. Agreements can be negotiated, with pastoralists being paid for engaging in actions which generate positive externalities or compensated for relinquishing rights which generate negative externalities.

Such multi-purpose land use is becoming more likely, particularly on the extensive arid and tropical margins and other areas with high conservation, landscape or recreational value. Leasehold tenure provides an appropriate mechanism for the evolution of multi-value landscape-specific systems of land use management, compatible with a primary land use, notably grazing.

The general principles underpinning these proposed lease tenures reforms are, firstly, to recognise the ongoing role of pastoralism as the primary use and, secondly, to require the pastoralist to adopt sustainable range management, but, thirdly, to ensure that other interests and values are adequately recognised within the 'bundle' of rights and duties attached to the lease, with prime attention to access and preservation of biodiversity. The fourth consideration is to ensure that option values for higher uses are adequately addressed, by providing mechanisms for the award of additional property rights. It is also important to ensure that the lessee cannot exercise monopoly control (or absolute veto rights) over conversion of land to more intensive uses.

All these provisions are consistent with a long-standing philosophy, underpinning lease tenure, to allow pastoralists to pursue their primary activity with minimum interference, but also to recognise that there are other interests meriting formal recognition. Pastoralists control disproportionately large tracts of land, to which they direct only modest inputs and generate comparably modest outputs. Given these low levels of economic return per unit area, the case for recognising other resource values is enhanced, particularly also given the semi-natural status of the rangelands. Furthermore, these specified constraints on the rights of lessees are consistent with the pragmatic institutional approach to property rights now endorsed by contemporary theory in resource economics.

3. Land tenure reform on marginal lands

Economic theory and practical experience both indicate a very different future allocation of property rights on the marginal lands. Significant property rights, in lease tenures, were awarded over much of these lands in the expectation that pastoralism was a viable broadacres use of these lands, capable of yielding an economic return through the customary inputs. However, due to an exceptionally adverse combination of environmental and locational disabilities, pastoralism has remained below the margin of profitability, save where operated as a cattle-harvesting rather than a cattle-husbandry operation (Johnston 1985, Holmes 1990).

The Brucellosis and Tuberculosis Eradication Campaign, when vigorously enforced, presented these marginal operators with a choice between two untenable courses of action — either engage in uneconomic capital investment to ensure effective herd management or remove the only significant asset, the cattle. Although the disease campaign was modified, to ease compliance in lightly stocked, seemingly disease-free areas, nevertheless marginal holdings in northern Cape York Peninsula, the Northern Territory Gulf District and the northern Kimberley either remain totally destocked or retain such a depleted number of cattle that they are having difficulty even in returning to their previous 'harvesting' mode of operation. For example, of the 26 cattle enterprises operating on the 30 pastoral leases covered in my 1985 Gulf District survey, 13 were found to be non-viable, under almost any set of assumptions, with these being located mainly on the coastal plain. Although BTEC has adopted a more tolerant approach to herd control in apparently disease-free

areas, nevertheless only one of the 13 leases is currently operating as a commercial enterprise, and this only with supplementary income from other sources. Two are now Aboriginal-owned with a third currently being purchased on behalf of Aboriginals. Two have recently been purchased by the Northern Territory government, for conservation and recreation purposes. Three are held by absentee owners whose intentions are not clear, while three others are held mainly for lifestyle reasons by owners whose place of residence shifts between Borroloola and the lease. The remaining lease is being subdivided, with half being held by a mining company and the other half by an occasionally resident owner. While these 13 enterprises encompassed over 40 000 km^2 or 38% of the area held in pastoral leases, they were estimated to have only 14% of the area's livestock carrying capacity, and only 9% of the managed-herd carrying capacity, comprising only 12 500 head of cattle (medium estimate).

There are similar trends in the other two most marginal regions with environmental and locational disabilities comparable to the Gulf District, namely Cape York Peninsula north of Coen and the northern Kimberley (Laut and Nanninga 1985). North of Coen, for example, there is currently only one fully commercial cattle enterprise. This is owned by Comalco and has a powerful locational advantage in supplying beef to the adjacent Weipa market. Three former pastoral leases have been converted to National Parks, while two are currently owned by the Queensland government, for non-pastoral purposes and one has been purchased for Aboriginal ownership.

In recognition of the incapacity of marginal leases to generate an adequate income, land administrations no longer enforce requirements for lessees to engage in pastoral enterprises. In the Northern Territory, lessees on marginal lands have been unable to attain the minimum standards of herd management and capital investment in order to achieve conversion to perpetual tenure. The Department of Lands and Housing has been grappling with the issue of an appropriate tenure for these non-viable leases, and is currently considering the merits of awarding crown leases. Normally this form of lease is to enable the lessee to engage in a specified enterprise, but this provision may need to be waived for lessees on marginal lands.

It is increasingly clear that there is a mismatch between land tenure and land use on the marginal lands. With the retreat of pastoralism, the case for private tenures, previously tenuous, has now disappeared. There is no other broadacres use in view, which requires secure private ownership of tracts, usually exceeding one thousand square kilometres, to enable private inputs of management, labour, capital or technology in order to generate income streams. While commercial tourism has been proposed as an alternative, the reality is that capital investment in tourist facilities is highly localised, involving only small land parcels by north Australian standards. The natural resources on which tourism is based.... the beaches, estuaries, rivers, rainforests, heritage sites and wildlife... are best treated as public goods, not merely because they have traditionally been used in this manner, but also because this is economically and socially more efficient. Any attempts to privatise these extensive natural assets will generate disbenefits to resource users quite disproportionate to the modest income to the 'resource owner', who would need to expend more effort in gathering fees than in providing services. Of course, where related services, such as accommodation and recreation facilities, are made available, they can well be privatised, but their land requirements can readily be met by licence, lease or freeholding of the land area needed for the facilities, within a matrix of public lands.

The disbenefits arising from overextension of private property rights beyond capabilities for effective resource use, have become increasingly evident. Outcomes have been consistent with those proposed in current theories in resource economics. Where resource rights are in excess of prospective resource use, there is a high propensity for the holders of these rights to engage in rent-seeking behaviour, directed towards capturing wealth rather than generating wealth. One form of rent-seeking behaviour is to charge entry fees without providing a commensurate level of services. A more newsworthy activity, of course, has been land speculation, particularly when it assumes epidemic proportions, as occurred in the late 1980s in Cape York Peninsula. Reports on Bromley, Bertiehaugh, Geike, Silver Plains, Starcke and other leases have been highlighted in the media. The underlying causes, and the outcomes have been discussed elsewhere (Holmes 1991) and need no repeating here, save to emphasize the negative outcomes, not only for the pastoral industry (because prospective pastoralists are forced out of the land market), but also for the inevitable, needed process of land reform, involving conversion of pastoral lease lands to other tenures.

In order to minimise this speculation, it is of critical importance that governments reaffirm two central principles of lease tenure: firstly that the lease is held essentially for pastoral purposes, and therefore that any compensation on resumption or payment on purchase will be on pastoral values (together with improvements); and, secondly, that the state has the right to acquire parcels of land for more intensive uses, including private uses. This second principle is to ensure that lessees do not engage in monopolistic behaviour through veto powers over other forms of development. Unfortunately, these two core attributes, historically central to the use of pastoral leases as effective policy instruments have been allowed to atrophy or even to be deleted in some jurisdictions. In Queensland, the Wolfe Report (Queensland 1990) recognised the problem and concurred with West Australia (1986)

on the desirability for the state to have the power to resume small parcels of land from pastoral leases for more intensive, private use.

4. Transfer of marginal lands to non-private tenures

In response to changing resource values, there is already a persistent transfer of marginal lands from pastoral leases to other uses and/or tenures. The largest single contribution to this shift has been by purchase for Aboriginal use. This is clearly revealed in the 1993 Australian Land Tenure Map, prepared by the Australian Surveying and Land Information Group, using official data as of May to July 1993. In the northern Kimberley district 20 006 km² of former pastoral leases have been transferred, comprising tenures currently titled: Carson River, Pantijan, Gibb River, Glen Hill, Mt. Barnett, Bow River and Doon Doon. In the Northern Territory, 15,306 km² have been acquired, including: Fitzroy, Eva Valley, Stokes Range, Hodgson Downs, Wada Wadalla and Garawa. More recently the well-known Elsey Station has also been purchased. In Queensland, 11 998 km² of lands in the northern savanna zone have been acquired, including: Morr Morr, Mission River, Helmsley, Ancilia, Pickersgill, Merepah and the Quinkan reserve, with additional areas from Starcke expected to be transferred. Bonny Glen and Kondaparinga Holdings in the Eastern Zone are also Aboriginal owned.

Most acquisitions have involved marginal leases, often run-down and destocked or with depleted herds as a result of BTEC. A prime consideration is the low purchase price (even though invariably well above market value), but also these lands generally have higher values to Aboriginal peoples than do the core grazing lands.

Over the last two decades, there have been substantial acquisitions for national parks and reserves, and this is likely to continue further, with one constraint being the need for relevant agencies to receive sufficient funds for operational purposes. Also some pastoral leases have reverted to ownership by state corporations or departments, now searching for some way of utilising these lands with some still held under lease title. Holdings in this category include Mataranka, Deepwater, St. Vidgeon, Nathan River and Billengarrah in the Northern Territory and Batavia Downs and Heathlands in Cape York Peninsula, with negotiations currently proceeding for the purchase of Starcke.

5. Public tenures to accommodate multiple uses

This process of land acquisition and reversion has so far occurred piecemeal. Although there has been a notable increase in lands held by government departments, no government has yet recognised that the changing resource context requires an expanding role for public ownership of these marginal lands. This will require new mechanisms to accommodate the resource needs of both Aboriginal and non-Aboriginal interests, within an emerging context of multiple values and interests. So far, only limited experience has been gained in this critical area, and largely confined to joint management of a few national parks involving Aboriginal interests.

It is worth recalling that Australian land administrators have previously had ample experience with flexible, adaptive instruments for awarding either property rights or use rights, through special-purpose leases, licences and permits. Certain time-honoured mechanisms such as annually renewable occupational licences with individually specified rights and responsibilities could well be ripe for revival. However, there is one critical difference between the previous use of flexible permits and titles and the emerging needs in the northern marginal lands. These instruments have historically been used in a piecemeal, *ad hoc* manner, to meet particular needs. In the new context, they will need to be used in a more comprehensive and purposeful way, both in achieving the needed transition in property rights and also in ensuring further adaptation to changing future resource challenges.

Currently, the nearest Australian land tenure model is the regional reserve title, now used in South Australia, particularly where negotiated as a successor to a pastoral lease, as with the Innamincka Regional Reserve. However, the tenure proposed here differs markedly from the Innamincka model in that the state clearly retains full title to the land. A more appropriate model is provided by the American federal lands, administered by the Bureau of Land Management and the National Forest Service. On these lands the challenges arising from multiple uses and conflicting values have been more sharply etched than in Australia (Holmes 1994b). An effective new model could be designed by selective choice of a blend of attributes from both American and Australian models of public tenures, where there is a capacity to award private use rights. During the transition, there will be a need to engage in purposeful negotiated trade-offs with existing titleholders, who may well be prepared to relinquish large tracts of 'useless' country (over which they are clearly made aware that they hold only grazing rights), for a package agreement which may possibly involve a mix of purchase payment and/or retention of a smaller land parcel but with expanded property rights and/or a part-time salary and/or priority rights to contract in order to supervise adjacent publicly-owned lands. Also, as stated earlier, there should be opportunities for Aboriginal involvement, in a variety of ways, in these new modes of resource use.

Within the portfolio of use rights available to private enterprises, there are likely to be expanded opportunities to award franchises and concessions, by auction or tender, where an appropriate income-earning activity can be identified and designated on public lands. It is important that existing local interests, including

Table 5: Precepts concerning Land Tenure of particular relevance to Tropical Lands Management

- Property rights should match resource use.
- Where net benefits are positive, resources should be privatised: equally, where net benefits become negative, adopt a policy of deprivatisation.
- Generally where an assured income stream can be gained through broadacres use and there are net benefits to internalisation, private ownership is preferable.
- Powers of resumption for more intensive uses should be available, but should be used only sparingly, and only where there is clearly a strong public interest to be served; rely primarily on voluntary sales.
- Pay compensation for the withdrawal of property rights; equally, require payment for the award of additional rights, through purchase price, rental or fee.
- Land Management is the responsibility of the landowner: if the owner cannot afford to manage, the case for private ownership is difficult to sustain.

Table 6: Land Tenure Reforms appropriate to Marginal Tropical Lands

- Where pastoralism is economically sustainable, retain pastoral lease tenures, but also develop streamlined procedures to award additional use rights geared to identified resource uses.
- Encourage additional uses complementary to pastoralism, with first right of refusal being available to pastoral titleholder.
- Where pastoralism cannot generate sufficient income for ongoing resource management, replace land titles based on exclusive occupancy rights with one based on non-exclusive access and use rights. This is best achieved under a new form of public land title.
- On public lands seek a match of access and use rights tied to resource values, with recognition of both aboriginal and non-Aboriginal interests.
- Encourage complementarity between public and private lands, including preference to local titleholders and other local people, in the award of use rights, such as franchises, licences and permits, as well as in tendering and employment on public lands.
- While landholders have an ongoing responsibility to meet basic costs in land management, they can appropriately receive public payments for external benefits, such as protecting valued habitats or species, or providing enhanced public access.
- Government should retain or re-acquire powers of resumption of small land parcels from leases where needed for more intensive uses. However, resumption powers should be used sparingly.
- Recognition can be given to the aspirations for private ownership of marginal lands, for reasons of preference, lifestyle or supplementary income, providing that the titleholders are prepared to accept full responsibility in meeting costs of land management and service delivery.

adjoining landholders, be granted opportunities to gain much-needed income through priority in the award of permits, licences or franchises, also including part-time salaries for undertaking a variety of work on public lands. In these ways, public landownership can enhance the local market economy in these remote regions.

It is hardly coincidental that, since this paper was written, a near-identical proposal has been presented in a broadly similar context, namely the woody weed infested lands of the Cobar pediplain (Robson 1994). This would involve voluntary relinquishment, by sale, of private titles to create a Multiple Use Reserve. This proposal is also made in a recognition of the inadequacy of income streams to private landholders. However, the Cobar proposal is driven by the current incapacity of landholders to manage their land conservatively, whereas the proposal for the tropical savannas is much more strongly directed towards effectively accommodating emerging demands for multiple use, not readily incorporated into broad-acres private titles.

It is not suggested that vast land tracts will need to be converted from pastoral lease to multi-purpose public ownership. Even on marginal lands, private pastoral, non-pastoral or multipurpose leases should be utilised where the net benefits of private title are clear. This will usually be the case where the primary use(s) can generate a viable income stream and where non-market values are less than market values. However, there are also large areas where the most significant resource values cannot be effectively privatised, the most noteworthy being the estuarine and riverine environments with which the tropical savannas are richly endowed. Adjoining lands are sufficiently extensive and of sufficient value for multiple uses that they merit recognition by a distinct form of public tenure, designed to accommodate flexible use, primarily towards serving non-market values (i.e. public goods), but also with a capability to award limited, selective rights for private commercial and non-commercial use, whether for graziers, safari tour operators, crocodile farmers or Aboriginal activities.

General precepts relating to land administration, together with policy-focused tenure reforms are recapitulated in Tables 5 and 6.

5. STRATEGIC REGIONAL PLANNING

During both the current transitional phase in restructuring tenure and also in the ongoing phase of flexible, multiple resource use, there is a clear need for comprehensive but adaptive regional land use and development planning. This need is being recognised, belatedly, at both national and state/territory levels, with a growing emphasis upon a broad-based, participatory approach, even in reports prepared by scientific organisations, including ASTEC (1993) and Australia (1993).

'There is an increasing trend towards landscape management on a regional basis. Such arrangements emphasize regional environmental characteristics, needs and responses and promote intergovernmental cooperation along with community participation.' (Australia 1993)

This need has already received some recognition, with preliminary planning exercises in the Kimberley, the Northern Territory Gulf District and Cape York Peninsula mentioned earlier. Consistent with the viewpoint pursued in this paper, governments have recognised that priority must be given to the marginal regions. I have reviewed these initial planning tasks, comparing them with current regional planning strategies in the Canadian Arctic which have been effectively reported and reviewed in Fenge and Rees (1987). Compared with northern Canada, planning in northern Australia is likely to have the following characteristics:

- 'There will be a lack of a system-wide, national approach, but separate evolution within the three relatively autonomous states/territory which, save in a few limited areas, retain the major powers relevant to effective land use planning.

- *Planning will be seen primarily as a bureaucratic task requiring technical expertise, with the political input to the planning task being mainly transmitted through higher-level decisions rather than direct, grassroots power. This is consistent with entrenched power-relations within Australian states.*

- *Centralised decision-making will be further demanded because of the larger number of special-purpose government departments with an interest in the process requiring higher-level determination.*

- *One further impetus to centralised decision-making arises from the entrenched local dualism between native and non-native interests in northern Australia, with these two sets of local interests often being so strongly divergent that the most critical regional decisions must inevitably be taken by the central government. While the dualism of local interests in northern Australia should not be used as an argument against substantial local participation, it is nevertheless likely to render this participation less influential in determining major outcomes.*

- *Accordingly, land use planning will mainly be perceived as a means of achieving the following limited goals: coordinated assembling and dissemination of information on regional resources and planning issues; some coordinated decision-making between the various responsible authorities; an enhanced decision-context through identification of priorities in land use and management for specific land tracts; a more constructive, informed and open response by these authorities to changing local circumstances; and a genuine attempt at effective land use and land management strategies for some public lands.*

- *The degree of interest and involvement by relevant state or territory governments will wax and wane according to the pressures for coordinated decision-making. Since the recent growth in such pressures is likely to be maintained in the immediate future, there will be a parallel governmental response.'* (Holmes 1992)

In my NARU paper (Holmes 1992), I attempt to make an appraisal of the Kimberley, Gulf and Peninsula planning projects, comparing their performance against a series of desired attributes, namely: well informed; dynamic and flexible; consultative and participatory; responsive and accountable; capable of resolving conflicts; co-ordinative; and influential. Furthermore, as emphasized by Rees (1987):

'Land use planning is neither an isolated process nor an end in itself. Ideally, land use matters will be an integral component of a more comprehensive planning process. When land use is considered within a framework of overall development policy, it becomes the spatial or geographic expression of regional social and economic objectives.'

The integration of land use planning within broader goals of regional economic and social development also enables the planning process to be used as an essential framework for the assessment of the environmental, economic and social impacts of development projects. In the tropical north all of these impact assessments, and especially social assessment, have become essential planning tools, given the disproportionate impacts of development projects on Aboriginal peoples. While these assessments are now being undertaken, as with Coronation Hill, Macarthur River and other projects, they will continue to prove unsatisfactory to all parties, while pursued in an isolated, *ad hoc* manner, divorced from a wider on-going, consultative regional planning process.

Social impact assessments, now being undertaken, are poorly linked to Aboriginal needs (Howitt 1993), and, indeed, the most striking deficiency of the three quoted regional planning efforts is the near-total failure to recognise and to seek to engage Aboriginal interests in the planning process. This raises serious questions about the validity of these planning efforts, given that Aborigines have the most substantial and durable land-related involvement in these regions, with an expanding land base and with near-equality between Aboriginal and non-Aboriginal populations. Aboriginals comprise 43% of the population of the Kimberley, 57% in the Gulf and 46% in Cape York Peninsula (55% if Weipa is excluded). The former total dominance by non-Aboriginal interests is now being replaced by a complex interplay of interests, posing a greater, but by no means insurmountable challenge in reconciling Aboriginal and non-Aboriginal interests. If pursued in a purposeful way, regional planning may well prove to be the most effective means for ensuring Aboriginal engagement in regional issues, which are of pivotal importance to their future. Unless these seemingly divergent interests are brought together at the local and regional level, the prospect is for deepening divisions in the north.

The current moves towards Aboriginal community development planning, sponsored by ATSIC are a move in the right direction and will help to develop awareness and capabilities among Aboriginal peoples

(Wolfe 1993). However, these efforts at community self-management should not be regarded as an alternative to Aboriginal participation in the wider regional planning task, involving a full engagement of Aboriginal and non-Aboriginal interests.

Given the very limited resources, short timespan and preliminary nature of the Kimberley and Gulf studies, as well as their limited terms of reference, it is hardly surprising that they fell short on most of the desired attributes, listed above. The currently proceeding Cape York Peninsula Land Use Strategy offers promise of better things but also gives cause for concern. On the positive side is an outwardly stronger governmental commitment, involving more substantial resources (nominally $9 000 000), over a longer time period (at least five years), with a shared involvement by state and federal governments, with full-time professional staff and with a very extensive programme of consultation and participation now being pursued through formally recognised community and interest groups, all actively engaged in the tasks delegated to working groups. This is leading to a constructive engagement of Aboriginal and Torres Strait Islander representatives with other interest groups, with positive responses from these representatives.

These are promising developments, following an inauspicious start. The core problem was that the most ambitious and expensive component in CYPLUS was announced and approved, even before any programme of consultation and participation was undertaken, thus placing the cart before the horse. The Natural Resources Analysis Programme, comprising Stage One, and now approaching completion, will consume almost half of the funding committed to CYPLUS. This programme can be criticised for a series of attributes, inappropriate to a regional research and planning programme.

A regionally-oriented research programme should only emerge following a consultative process seeking to identify regional problems and potentials, leading to decisions on research priorities. This programme was announced without effective grassroots consultation. It is hardly surprising that local groups, particularly Aborigines, became alienated at an early stage.

While all seventeen projects in Stage One clearly have merit and do cover a wide array of pivotal issues, nevertheless they comprise a series of discrete projects, with little opportunity for the cross-disciplinary fertilisation which should be a central strength in contemporary regionally-focused research. In particular, there is a conspicuous lack of involvement of social scientists, and the programme is thereby severely weakened. I have already expressed my astonishment that the recent ASTEC report (1993) on the role of the social sciences in research and development should quote CYPLUS as an exemplary model of multidisciplinary research being together the natural and the social sciences. In my view, it should be cited for precisely the opposite reasons!

Effective regional research can only emerge through an initial intense exchange of ideas between the main interest groups and an informed, multidisciplinary team of researchers, preferably all knowledgeable about the region. This offers the best means for ensuring well focused, reasonably integrated research programme, in which the whole is more than just the sum of the parts. This is essential where large numbers of research groups are involved. In more modest, regionally-focused, multidisciplinary research, it may be possible for one or two researchers to achieve the necessary synthesis, particularly when focused on a set of clearly defined, closely related issues, such as undertaking land appraisal in order to assess grazing capabilities and land use options in a comparable marginal region (Holmes 1990). If many research teams are to be engaged in a regionally-focused research programme, then substantial preliminary work is essential.

CYPLUS, Stage One, will undoubtedly yield valuable contributions, particularly in furthering disciplinary-focused knowledge. However, from a regional viewpoint its main function will be as a useful information source rather than as a well-focused contribution directed towards the Peninsula's problems and potentials and furthering the challenging task of identifying appropriate development strategies. (Less my strictures on CYPLUS research should be interpreted as those from a disappointed applicant, I hasten to add that, at the time of writing this paper, I have had no involvement in any CYPLUS-related research applications).

Given the lack of experience of all participants with the culture and procedures for regional strategic planning, it is to be expected that initial outcomes will be disappointing, perhaps even frustrating, to all involved. It will be helpful if the first few years are treated as a learning experience, in the expectation that steady persistence will yield substantial benefits in due course. Given the mounting challenges, specific to these northern marginal regions, the absence of co-ordinated, long-term planning will only result in fragmented, often conflicting policies and programmes, with very adverse outcomes.

6. Conclusion

In Australia's tropical savannas, there are other major issues ripe for substantial institutional reform, not addressed in this paper. One of these, close to my own interests, relates to the restructuring of settlement and servicing systems to mitigate the problems arising from a sparse, dispersed population, scattered over a vast area. A further topic, listed for this talk, concerns ecologically sustainable development. While I regard these as important matters, I see little purpose in discussing them until the most basic areas for institu-

tional reform are addressed. In this paper, I have sought to demonstrate that the two most fundamental aspects are, firstly, reform of property rights systems to ensure that they are compatible with the emerging resource context, and, secondly, implementation of regional strategic planning, in a context where a more purposeful, co-ordinated approach is essential, if the formidable challenges are to be properly understood and effectively tackled. A necessary precondition is a more informed understanding of the pronounced shifts in savanna resource values currently being experienced, and how these are linked to powerful, ongoing forces for social and economic change.

Acknowledgments

Preparation of this paper has been considerably assisted by the ready cooperation of people within the W.A. Pastoral Board, N.T. Department of Lands and Housing and Queensland Department of Lands. Research into land tenures has benefitted from discussions with many people in all states and territories. This has been acknowledged in other papers. On matters specific to this paper, I am grateful for help and advice received from: Peter Jull, Lou Kelly, Laurie Knight, John Mott, Gary Swanson and Mike Young. However, responsibility for the views expressed rests solely with the author. Over the period 1991-1993, research on pastoral land tenures was supported by grants from the Australian Research Council. This paper has been written while the author was Research Associate in the Environment and Behaviour Program, Institute of Behavioural Science, University of Colorado, Boulder. I am particularly grateful to the office staff, and particularly to Barbara Marshall, in the Department of Geographical Sciences and Planning at The University of Queensland for their forbearance in processing this script by long-distance contact.

References

AUSLIG (Australian Surveying and Land Information Group) (1993). *Map of Australian Land Tenure*. Government Printer, Canberra.

Australia. Office of the Chief Scientist (1993). *Research and Technology in Tropical Australia: Selected Issues*. AGPS, Canberra.

ASTEC (Australian Science and Technology Council) (1993). *Research and Technology in Tropical Australia and their Application to the Development of the Region: Final Report*. AGPS, Canberra.

Bridgewater, P.B. and Walton, D.W. (1995). Tropical Savannas: National Parks and other protected areas. In: *The Future of Tropical Savannas: An Australian Perspective*, (Ed. Ash, A.J.) pp. 54–61. CSIRO Publishing, Melbourne.

Bromley, D. (1991). *Environment and Economy: Property Rights and Public Policy*. Blackwell, Oxford.

Fenge, T. and Rees, W.E. (Eds.) (1987) *Hinterland or Homeland? LandUse Planning in Northern Canada*. Canadian Artic Resource Committee, Ottawa.

Holmes, J.H. (1990). Ricardo revisited: submarginal land and non-viable cattle enterprises in the Northern Territory Gulf District. *Journal of Rural Studies*, 6: 45–65.

Holmes, J.H. (1991). Land tenures in the Australian Pastoral Zone: a critical appraisal. In: *North Australian Research: Some Past Themes and New Directions*, (Eds. Moffat, I. and Webb, A.), pp. 41–59. NARU, Darwin.

Holmes, J.H. (1992). Strategic regional planning on the Northern frontiers. Discussion Paper 4. NARU, Darwin.

Holmes, J.H. (1994a). Changing rangeland resource values: implications for land tenure and rural settlement. In: *Outlook 94, Vol. 2, Natural Resources*, pp. 160–195. Australian Bureau of Agricultural and Resource Economics, Canberra.

Holmes, J.H. (1994b). Land tenures, property rights and multiple land use: issues for American and Antipodean rangelands. In: *The Place of Geography*, (Eds. Cliff, A.D., ld, P., Hoare, A.G. and Thrift, M.) Blackwell, London.

Holmes, J.H. (1995). Regional Restructuring of the tropical savanna: Impacts on lands, peoples and human settlements. *The Future of Tropical Savannas: An Australian Perspective*, (Ed. Ash, A.J.) pp. 5–19. CSIRO Publishing, Melbourne.

Holmes, J.H. and Knight, L. (1994). Pastoral lease tenure in Australia: historical relic contemporary tool? *The Rangeland Journal*, 16: 106–121.

Howitt, R. (1993). Social impact assessment as 'applied peoples' geography. *Australian Geographical Studies*, 31: 127–140.

Johnston, J. (1985). Factors affecting financial viability of remote area beef properties. In: *Brucellosis and Tuberculosis Eradication Campaign Workshop*. Department of Primary Industries, Brisbane.

Laut, P. and Nanninga, P.M. (1985). *Landscape Data for Herd Disease Eradication in Northern Australia*, CSIRO Division of Land and Water Resources, Canberra.

Morton, S.R., Stafford Smith, D.M., Friedel, M.H., Griffin, G.F. and Gillison, A. (1985). Australian savanna ecosystems. In: *Ecology and Mangement of the World's Savannas*, (Eds. Tothill, J.C. and Mott, J.J.), pp. 56–82. Australian Academey of Science, Canberra.

Mott, J.J., Williams, J., Andrew, M.H. and Gillison, A. (1985). Australian savanna ecosystems. In: *Ecology and Management of the World's Savannas*, (Eds. Tothill, J.C. and Mott, J.J.), pp. 56-82. Australian Academy of Science, Canberra.

Northern Territory Department of Lands and Housing (1991). *Gulf region Land Use and Development Study 1991*. N.T. Department of Lands and Housing, Darwin.

Queensland Land Use Policy and Administrative Review Committee (1990). *Report of a Review of Land Policy and Administration in Queensland*. Governement Printer, Brisbane.

Rees, W.E. (1987). Introduction: a rationale for land-use planning. In: *Hinterland or Homeland? Land Use Planning in Northern Canada*, (Eds. Fenge, T. and rees, W.E.) Canadian Artic Resource Committee, Ottawa.

Robson, A.D. (1994). The case for multiple-use reserves in woody weed country. Working Papers: Eigth Biennial Conference, Australian Rangeland Society, June 1994, pp. 58–61.

Western Australia, Department of Regional Development and the North West and Department of Planning and Urban Development (1990). Kimberley Region Plan Study Report: A Strategy for Growth and Conservation.

Wolfe, J.S. (1993). *Regional planning by ASTIC Councils: Purposes, Process, Product and Problems*. Discussion Paper 18. NARU, Darwin.

Savanna Users and Their Perspectives

Chapter 4

Savanna Users and Their Perspectives: Grazing industry

John W. Stewart

Glenlyon Pastoral Management Pty Ltd,
PO Box 1011,
Townsville, QLD 4810
Australia

Abstract

The grazing industries (particularly beef cattle) have been major users of the north Australian savannas for over a century. Permanent settlement commenced soon after the first explorers traversed the country and in many areas the grazing industries have been the only users.

The development of the grazing industries has been dominated by the harsh environment in the savannas (for both humans and animals) and the lack of local markets. The beef industry before 1950 was based on British breed animals grazed on native pastures with generally low production levels. Since then there have been many changes — the use of Brahman cattle, increased infrastructure on properties, supplementary feeding, improved animal health, sown pastures — and production has markedly increased. New markets were developed in USA and more recently in Asia.

Research has played a major role in improving productivity in the beef industry and will continue to be a priority for northern Australia. Producer and government contributions to research are managed by the Meat Research Corporation and the recently established North Australia Beef Research Council will have an increasing role in research and development.

Current major concerns of the beef industry are the need for ecologically sustainable development, government charges and provision of services, and land tenure. Important trends in the future will be increased markets in Asia, higher quality beef meeting defined product specifications, greater community pressure for involvement in issues related to the beef industry, continued improvements in production, changes to reduce costs in the processing sector, and pressure to maintain research capability.

Current indications are that the future market for both live cattle and beef will be buoyant; the challenge for producers is to sustain their properties over a wide range of seasons and to protect their financial viability by containing costs.

1. INTRODUCTION

Northern Australia covers an area of 3.5 million km^2 (1.2 million km^2 in northern Queensland, 1.3 million km^2 in the Northern Territory and 0.9 km^2 in northern Western Australia) which is 46% of the area of Australia. Tropical savannas cover over 80% of this area, the remainder being the wet tropical coast and the deserts of central Australia. The human population in northern Australia is sparse with current numbers being about 950 000 (ASTEC 1993); much of the population is concentrated in northern Queensland east of the Great Dividing Range and only about 200 000 people live in the savannas.

The grazing industry has been a major user of the tropical savannas for over a century. Cattle predominate but sheep (for wool) are also important. Grazing commenced in the south-eastern parts in the 1850s and then expanded rapidly so that most of the area was occupied by 1900. The grazing industry remains the major land use for large areas though the area used for pastoralism is declining (see Chapter 3 this volume). In this paper I examine the grazing industry in northern Australian savannas in terms of history, present status, and future opportunities.

1.1 History (pre-1950)

The first livestock were introduced to the early settlements in the Northern Territory (Melville Island in 1824, Raffles Bay in 1827 and Port Essington in 1835). When these failed the livestock were released and a few herds survived in the wild, including the Banteng cattle on Coburg Peninsula near Port Essington.

Between 1840 and 1880 explorers (e.g. Leichhardt, Gregory in the Victoria River district and north Queensland, Stuart in the Northern Territory on a number of trips, searchers for Burke and Wills, Forrest in the Kimberleys) gave glowing, but often inaccurate, reports of the pastoral potential of northern Australia and overlanders followed closely after. Settlement commenced in the south-east of the region in the 1850s and moved to the north and west and into the Northern Territory in the 1860s and 1870s. The east to west movement continued into Western Australia in the 1880s (e.g. Buchanan, McDonalds, and Duracks) and parts of the Pilbara and Kimberley were occupied from the south-west during this time.

The initial settlement was often a mix of sheep and cattle. Wool was both a more profitable and more durable product and cattle were pushed to frontiers where sheep were not viable due to problems with penetrating grass seeds, dingo attacks, fleece rot, foot rot, scab and long distances between water points. Apart from the Mitchell grass region of western Queensland (8 million sheep) and parts of the Hammersley and Pilbara area (300 000 sheep), the beef industry became the major, and sometimes only industry throughout the tropical savannas. The remainder of this paper will deal with the beef industry although some aspects and comments are also relevant to the use of savannas for wool production.

Two factors have dominated the development of the grazing industry: the harsh environment and a lack of local markets. High temperatures, a humid wet season and long dry season provide stress to both humans and animals. Herbage is of poor quality due to low fertility soils and rapid maturation and lignification of pasture during the hot, wet summers. The quality of herbage, already poor by the end of the wet season continues to decline during the long dry season. Consequently, herbage quality, and sometimes quantity limits animal production for much of the year ('t Mannetje 1982) and methods to improve animal nutrition have occupied producers and researchers since the industry developed.

At settlement, the immediate task was to get stock on the country but the search for markets soon became more important. The initial reliance on a small domestic market meant restricted sales, although gold discoveries and the subsequent population growth gave short term boosts in some areas (e.g. Charters Towers 1870, Palmer River 1876, Halls Creek 1886). Live shipments from the north west first took place during the 1880s and activity in this area has fluctuated ever since in response to changing profitability. Cattle ticks and associated tick fever were first recorded in the 1880s. These caused heavy losses, added to expenses and hampered stock movements, and still affect the industry today.

1.2 Structure of the cattle industry

Land ownership was spread over three groups: large companies, families with more than one property, and single family units. Property size varied widely. In Western Australia the maximum area for one lease was 1500 sq miles (3885 km^2); in the Northern Territory there were (and still are) 240 pastoral leases with an average size of 2000 sq miles (5180 km^2). There was much greater variation in Queensland where properties ranged from 50 sq miles (130 km²) to 2000 sq miles (5180 km²). Aboriginal stockmen were a majority of the labour force in Western Australia, the Northern Territory and northern Queensland, and a smaller proportion in central Queensland.

Herds were initially British breeds, mainly Shorthorns and Herefords with Shorthorns being the most prolific. Devons crossed with Shorthorns were also in demand. In most of the savannas, harvesting of stock, rather than managed herds, was the usual pattern. Bullocks in good condition dressed out at 220 kg at slaughter at 5- to 6-years-old and cows at 150 kg.

There were few fences, yards or equipped watering points. Drought, flood, fire and ticks made the life of the cattleman difficult. Natural herbage, mainly grasses, provided the feed for the cattle. The species

varied across the region with black spear grass (*Heteropogon contortus*), kangaroo grass (*Themeda triandra*), *Aristida* spp., *Bothriochloa* spp., annual sorghum (*Sorghum* spp.), Mitchell grass (*Astrebla* spp.), Flinders grass (*Iseilema* spp.), Queensland blue grass (*Dichanthium sericeum*) and spinifex (*Triodia* spp., *Plectrachne* spp.), being some of the major species.

1.3 Changes: 1950–1990

The situation began to change after World War II. Infrastructure began to improve, particularly with the sinking of dams and the provision of artesian and subartesian bores. Buffel grass (*Cenchrus ciliaris*) was sown in a number of areas to improve production but it was not until the 1960s that the big changes occurred.

Many things happened together. Brahman cattle were increasingly used: in Queensland in the mid-1960s less than 10% of cattle had some Brahman content but this increased to 80% in 1990 and more than 90% of properties in the savanna zone now have Brahman and Brahman-cross cattle (O'Rourke *et al.* 1992). The greater resistance to heat stress, stronger resistance to cattle ticks, increased foraging ability and higher forage intake of the Brahman cattle (Frisch and Vercoe 1977, Siebert 1982) improved the production and the survival rates of northern cattle.

Supplementary feeding of cattle increased markedly, using a wide range of supplements but particularly urea and phosphorus in various forms. Supplements increase the intake of low quality forage by stock (Coates 1995) and improve their survival, growth and calving performance (Winks 1990).

Animal health improved, particularly with the campaign to wipe out pleuro-pneumonia. This operation exceeded all the expectations of the planners. The availability of botulism vaccines enabled deaths to be markedly reduced.

In 1970 the Federal government established the Brucellosis and Tuberculosis Eradication Campaign (BTEC). This campaign, which saw the country rid of brucellosis in 1988 and tuberculosis in 1992, created a big infrastructure change in the northern industry. Provision of paddocks, yards and water points were paramount to the progress and success of the eradication campaign. General management of properties improved, weaning became the norm, and the use of better bulls saw a sustained improvement in both the type and productive capacity of cattle.

Research and development played a large part in the progress as calving performance, growth and survival all improved due to better nutrition, disease control and cattle selection. The use of introduced grasses (mainly buffel) and the introduction of Townsville stylo (*Stylosanthes humilis*) both increased. New stylo cultivars (shrubby stylo (*S. scabra*) and Verano (*S. hamata*)) were released to replace Townsville stylo which was decimated by anthracnose.

With all these advances in management, disease control and technology the days of the feral herd are gone. While BTEC has had many detractors it has prepared the northern industry for the 21st century. No longer do 5- and 6-year-old bullocks go to the meatworks; they are more likely to reach slaughter weights as 3-year-olds.

In terms of the beef trade, there was a big change in markets in the late 1950s and 1960s. Prior to that time, the bulk of the exports were to Europe, particularly the United Kingdom. The USA then entered the market looking for manufacturing grade beef. This was a boon to the cattle industry as it gave many cattlemen the opportunity to stop culling for age cows. USA soon became the biggest market for Australian beef and remained in that position until recently when Japan replaced it.

2. THE PRESENT

Since the 1950s property sizes in Western Australia (WA) and the Northern Territory (NT) have remained static. However in Queensland, where closer settlement was a goal of government during the 1950s and 1960s, many properties were reduced in area. As the leases on large properties expired the owners were left with one or two living areas (the size of a property considered necessary to make a reasonable living) and the balance was distributed by ballot. The areas most affected were north of Charters Towers and the Brigalow region of central Queensland. A survey (O'Rourke *et al.* 1992) of northern Australian beef producers in 1990 found property sizes in the savanna regions ranged from an average of 30 000 ha in the speargrass zone to 370 000 ha in the NT/WA spinifex zone. Herd sizes averaged 2000 head in the more developed areas and up to 9000 head elsewhere. Stocking rates also varied widely from less than 10 to more than 100 ha/beast. Virtually all properties had breeding herds with many also running fattening operations.

The nature of property ownership has changed with an increase in the number of single family units. A number of large pastoral companies have withdrawn and sold their properties. Many of the sales have been to other companies or to families with more than one property. In the Northern Territory and Kimberley, Aboriginal ownership of properties has increased markedly with Aboriginal interests now controlling 43 properties (see Pearce, Chapter 9 this volume). While Aboriginal ownership has increased, Aboriginal labour on properties has declined and Aborigines now make up only a small proportion of the workforce.

In the early 1970s the cattle market collapsed as a result of over-production and matters beyond Australia's control. With low prices, sales were reduced. This,

Table 1: Exports of live cattle from Darwin, Karumba, Townsville, and ports in northern Western Australia (AMLC pers comm).

Year	Breeding	Slaughter	Fattening	Total
1991	22 892	13 960	44 097	80 949
1992	26 522	18 268	78 281	123 071
1993	12 627	17 390	142 595	172 612

combined with good seasons in the mid-1970s contributed to the Australian cattle herd reaching a peak of 33 million head in 1976 (Australian Bureau of Statistics 1978). Through the 1980s and into the 1990s the herd remained at somewhere between 23–24 million head. Drought has been a particular constriction in many of the main cattle growing areas in the 1990s ensuring that the national herd has remained static.

At present about 7.5 million of Australia's 23 million beef cattle are grazed in northern Australia with 5.5 million in northern Queensland, 1.4 million in the Northern Territory, and about 0.8 million in northern Western Australia (AMLC 1993). The majority of these cattle are in the savanna areas. Northern Australia produces 18% of Australian beef exports and 22% of Australian beef exports to the USA. Cattle turnoff from northern Queensland is 520 172 head, 19% of the total Queensland turnoff of 2 704 020 and an estimated 5% of total cattle turnoff in Australia (Australian Bureau of Statistics 1993).

Exports of beef and veal from northern Australia have declined in the past couple of years because of continuing drought conditions. In contrast, exports of live cattle have increased markedly, particularly feeder steers for fattening (Table 1). This increase has proved to be a boon to cattle producers in the tropical savannas, helping to improve viability.

Today, prices for cattle in the northern cattle industry are buoyant. For the first time, savanna cattlemen have three groups interested in purchasing their store steers: live exporters, feedlotters and grass fatteners. This will lead to better returns for northern producers. The continuing trend to better quality cattle that can be finished at a younger age, increased feedlotting to meet the requirements of the export market and the domestic trade, increased live exports from the north, and the increasing use of stylo pastures to fatten cattle, are all adding value to the industry.

3. INDUSTRY CONCERNS

Current major concerns of the industry are sustainable production, government charges and provision of services and land tenure.

Ecologically sustainable development

There are three over-riding issues confronting the northern beef industry: the need to reduce and control costs while maintaining or enhancing output; the development of economically, environmentally and socially sustainable production systems; and the need to exclude and control diseases and pests (including woody weeds).

Producers have always been aware of the necessity to protect their land and they are more aware than ever of the need to monitor production performance in the light of sustainability. Degradation has occurred where financial constraints have made concentration on short term returns rather than long term persistence imperative for survival, where producers lacked knowledge of new areas, methods and techniques, and sometimes where producers have been driven by short term exploitation. Drought and woody weeds have much to do with the initiation of degradation. Conversely, during a good wet season, productive vegetation can reestablish on country considered to be degraded.

Woody weeds pose a major and increasing threat to the future of the beef industry in the savannas. In addition to native species which are increasing in density in many areas, there are several introduced species now widely established in northern Australia. These include prickly acacia (*Acacia nilotica*) which is particularly prominent in the Mitchell grass region of Queensland, rubber vine (*Cryptostegia grandiflora*) in many of the river systems of north-east Queensland, as well as mesquite (*Prosopis* spp.), chinee apple (*Ziziphus mauritiana*), and parkinsonia (*Parkinsonia aculeata*). Dense stands of these species reduce herbage production and carrying capacity. Ease of access and mobility are also reduced, making stock handling more difficult.

One of the main driving forces in tackling these issues of sustainability has been the Landcare movement, which was initiated by rural producers for rural producers. Landcare has exceeded all expectations in the effect it has had on changing attitudes and behaviour, not only of rural landholders but the community in general.

Much of the 'on-farm' research now being carried out in the savannas is focussed on maintaining or improving sustainable production. A goal of the Meat Research Corporation's (MRC) North Australia Program 2 is '*to have 20% of the degradation-prone land in northern Australia operating under land use production systems which are environmentally sustainable and commercially viable by the year 2000*' (MRC 1993).

Various state governments have introduced property management planning as a tool to assist producers in achieving sustainable development. While this development has been valuable in providing a better resource information base to producers, I and a number of others have concerns that this process is giving governments too much information to base their future decisions on, which could be detrimental to land holders.

More recently, the Commonwealth government formed The National Rangeland Management Working Group (1994) to develop a National Strategy for Rangeland Management. Commitment to develop such a Strategy arose out of the Environment Policy Statement by the Prime Minister in March 1993. This statement committed the Commonwealth to work cooperatively with State and Territory governments, traditional owners, industry, the farming community and conservation groups to develop a national strategy, including an action plan, for the ecologically sustainable use and conservation of Australia's rangelands.

With all these measures in place I consider sustainability is being addressed. From an animal or animal welfare view the industry has been able to answer the critics and produce its own welfare plans for feedlots, transport and on-property procedures.

Government charges and provision of services

Governments can have a huge effect on the cattle industry through their control of rents and taxes, both of which can make it difficult for producers to remain viable. The 'user pays' system is high on the agenda of all governments. The industry accepts this provided that the 'community' share is met particularly in areas that assure our 'clean, green product'. If governments impose more costs on the industry then industry needs to ensure that it has a substantial say in how the funds raised are spent.

Due to cutbacks in funding for government animal health services right around Australia (to allow more dollars for social welfare, health and education), it is necessary to set up a body, which will probably be known as the National Animal Health Commission, to restore animal health services, and through a National Animal Health Information Service, ensure the acceptability of our product and live exports to importing countries. This will possibly be funded equally by the Commonwealth, the States, and industry.

Across northern Australia, government services have become regionalised. This has generally meant a cutback in services because the local stock inspector or the Clerk of Petty Sessions is no longer part of the local township. The service is perhaps still there but has to come from the regional centre. It is supposed to be more cost effective but I wonder.

It is essential that the industry lobby governments to ensure that their policies are more favourable to assisting rather than hindering the on property operations of the cattle industry.

Land Tenure

Almost all the land used for cattle grazing in northern Australia is leased from the various state governments under a wide range of leases with different covenants and restrictions, e.g. in the Kimberley/Pilbara region of Western Australia the maximum area for a lease is 3885 km^2. In the Northern Territory, a new Pastoral Land Board established to administer the land is examining various land titles. Security of tenure has long been a concern within the industry and the uncertainties have been increased in recent times following the Mabo decision in the High Court in 1992. There is a concern in the industry that some of the good cattle country may become unproductive with aboriginal ownership. This is a trend already. It must be reversed.

Research

The cattle industry thrives through changes made available by research. Results from research on cattle management (weaning, use of improved bulls, culling for fertility, crossbreeding) and nutrition (supplementary feeding, improved pastures) have changed the industry. Research will continue to be a priority for northern Australia to ensure that we are able to meet the developing markets in southeast Asia and maintain our exports to existing markets.

Research for the cattle industry is managed by the Meat Research Corporation which currently raises $48 million per year; $24 million from the producer/processor levy on each beast traded, and $24 million from the Commonwealth Government based on their commitment to fund up to 0.5% of the GVP (Gross Value-added Product) of the industry.

The North Australia Beef Research Council (NABRC) arose out of a meeting of beef producers, researchers and funding bodies interested in ensuring that research for the northern beef industry dealt with problems identified by that industry. Seven regional committees across Queensland, Northern Territory and northern Western Australia are represented on the Council which commenced operation in November 1992. It takes the broad view that research includes all the functions for success: research, development, extension, education and training.

Producers, feedlotters and processors are represented on NABRC, as are research providers and funding bodies including CSIRO, state Departments of Primary Industries and Agriculture, universities, and the Livestock and Meat Authority of Queensland. An observer of the Meat Research Corporation also attends Council meetings, so for the first time all players in cattle industry research are represented in one body.

The area represented by the NABRC covers nearly 50% of the cattle population of Australia, representing 11.3 million head. It is absolutely essential that research in the cattle industry be directed to maintain current markets and ensure full advantage is taken of developing markets in southeast Asia for both beef and live cattle for breeding and feeding.

With its regional structure, the NABRC is in an ideal position to collate and develop issues that cattlemen see as necessary to improve production in a sustainable manner, while exploring all avenues to reduce production costs. Because the NABRC is producer driven, it is considered that through contact with producers, the Council and its Regional Committees will enhance technology transfer through existing extension mechanisms. The Council intends to establish and maintain a pool of trained personnel who are available at all times to assist in decision making in the areas of research, development and extension for the cattle industry.

Recently, the NABRC (1994) produced its Strategic Plan. This plan, together with regional plans, will become the blueprint for research and development for the north Australian cattle industry into the 21st century. It will, of course, be responsive to the changing needs of the northern industry.

The Future

The latest Australian Bureau of Agricultural and Resource Economics (ABARE 1994) forecasts are:

- the Australian saleyard indicator price for cattle is forecast to increase from 238 c/kg in 1993-94 to 280 c/kg in 1994/95;
- beef production in Australia is forecast to fall marginally in 1994-95 to 1.71 million tonnes;
- significant increases in market access for Australian beef are expected as a result of the successful conclusion of the Uruguay Round of GATT;
- the herd is estimated to have increased to 23.8 million head in March 1994;
- beef consumption is forecast to fall to under 35 kg per person in 1994/95;
- there is an expected retail price rise of 6% in 1994/95.

The North Australia Beef Research Council sees major trends in the following areas over the next 15–20 years; markets, community influence, production, processing, research, political influence, and others.

Markets

Market prospects are good with new opportunities arising provided Australia can continue to service current markets and open new ones. The current overseas focus will increase, particularly in Asia as the average income of the population improves. Consumption of beef is expected to rise, increasing the requirement for live cattle; the live cattle trade is expected to double in the next ten years. Both local and overseas markets will have more clearly defined product specifications, the quality will be higher, new products will become available, residues will be lower and the product will be increasingly safe. Industry is concerned with the uncertainty of exchange rates as an increase in the value of the dollar reduces the return to the industry. Competition from subsidised products will continue to be a problem and there needs to be more emphasis on value-added meat products.

Community Influence

In recent years there has been increased community pressure and demand for involvement in issues such as Aboriginal land rights, land access and tenure, animal welfare, and conservation/landcare. The industry must ensure that the wider community has a better perception of agriculture and that governments are aware of industry concern at declining rural infrastructure and population partly due to the continuing removal of services throughout rural Australia. The current trend towards an ageing producer population needs reversing and we need increasing education (including tertiary level) for future cattlemen.

Factors influencing production

There will be a continued drive to produce high quality cattle at a younger age. In savanna areas suitable for improved pasture there will be an increase in the area sown. There is a huge area of the savannas in higher rainfall areas with the potential to be developed. In other areas, particularly marginal grazing lands, carrying capacities will be reviewed and in some cases with a reduction in stock numbers, better returns will be achieved. If a tick vaccine becomes a reality I see a number of producers crossing Brahmans with British and European bred cattle to produce a more suitable product for the feedlot.

Improvements in production will come from improved breeding technology to produce better cattle, more sustainable on-farm management practices (property management planning and landcare will play an increasing role), and management of woody weeds (or biological control if it becomes a reality). The National Rangeland Management Strategy will have an impact as will a Co-operative Research Centre in Tropical Savannas if it becomes a reality in this current round for Co-operative Research Centres. Exotic diseases will remain a threat, especially with increased tourist trade, boat people and possible declining quarantine services, while the seasonal nature of supply and dry season downturn in production in northern Australia will continue to be a problem.

Processing infrastructure

The processing side of the industry must address the cost of processing in this country. It must make every effort to reduce costs in this area by increasing production in meatworks with infrastructure changes, including perhaps greater use of automated killing facilities. It is likely there will be fewer processing facilities in the future.

Research Capability

Research will provide better cattle and an improved product, which in turn will improve income. Research

is also necessary to ensure the industry becomes more cost effective and so contain property running costs. It is extremely important for the future of the industry that the research potential and quality of research providers is maintained. With continued pressure on governments there is a risk of decreased investment in research and political influence will be required to at least maintain the *status quo*.

Political influence
As in other areas of agriculture, the number of people directly involved in the beef industry is decreasing and their political influence is declining. The industry must work to make inputs into the agenda of national and state governments, and ensure their policies are supported by government.

The availability of rural finance is declining due to both political and commercial decisions. There has been much criticism of banks and their attitude to rural borrowers. Ours is a long term industry. Seasons can have such a detrimental effect on viability that banks must be in it for the long haul.

Other influences
There are a number of other factors which will influence the future of the industry: climate change (greenhouse effect, global warming, general change to rainfall); corporation and stratification (changes in company and private ownership); better infrastructure (transport, improved roads, mail, etc.); and alternative enterprises (e.g. mining, tourism).

4. CONCLUSIONS

A viable cattle industry in the savannas should continue. It is the only industry providing some population and infrastructure in much of this huge area. Mining, while supplying people, tends to be congregated in small areas and there is an increasing trend of fly-in-fly-out operations, which further lessens the socioeconomic influence of mining in the savannas. Major tourist activities are also concentrated in restricted areas. There is every indication that the future market for both live cattle and beef will be buoyant. The concern for producers is the need to plan for sustainability of their properties over a wide range of seasons and to protect their financial viability by containing costs.

References

ABARE (1994). Beef and veal. *Australian Commodities: Forecasts and Issues*, **1**: 17-18.

Australian Bureau of Statistics (1978). *Year Book Australia. No. 62, 1977 and 1978*. Australian Bureau of Statistics, Canberra.

Australian Bureau of Statistics (1993). *Livestock and Livestock Products, Queensland 1992-93*. Australian Bureau of Statistics, Brisbane.

AMLC (Australian Meat and Livestock Corporation) (1993). *Statistical Review. June 1992-July 1993*. Australian Meat and Livestock Corporation, Sydney.

ASTEC (Australian Science and Technology Council) (1993). *Research and Technology in Tropical Australia and their Application to the Development of the Region. Draft Report.* AGPS, Canberra.

Coates, D.B. (1995). Effect of phosphorus as fertiliser or supplement on the forage intake of heifers grazing stylo-based pastures. *Australian Journal of Experimental Agriculture*, **35**: (in press).

Frisch, J.E. and Vercoe, J.E. (1977). Food intake, eating rate, weight gains, metabolic rate and efficiency of feed utilization in *Bos taurus* and *Bos indicus* crossbred cattle. *Animal Production*, **25**: 343-358.

Mannetje, L. 't (1982). Problems of animal production from tropical pastures. In: *NutritionalLlimits to Animal Production from Pastures* (ed Hacker, J.B.), pp. 67-85. Commonwealth Agricultural Bureaux, Farnham Royal, UK.

Meat Research Corporation (1993). *Project Guide. 1992-1993*. Meat Research Corporation, Sydney.

National Rangeland Management Working Group (1994). *Rangeland Issues Paper, February 1994*. National Rangeland Management Working Group, Canberra.

North Australia Beef Research Council (1994). *Strategic Plan. June 1994-2000*. North Australia Beef Research Council, Brisbane.

O'Rourke, P.K., Winks, L. and Kelly, A.M. (1992). *North Australia Beef Producer Survey 1990*. Queensland Department of Primary Industries, Brisbane and Meat Research Corporation, Sydney.

Siebert, B.D. (1982). Research findings in relation to future needs. *Proceedings of the Australian Society of Animal Production*, **14**: 191-196

Winks, L., (1990) Phosphorus and beef production in Northern Australia. 2. Responses to phosphorus by ruminants — a review. *Tropical Grasslands*, **24**: 140–158.

Chapter 5

National Parks and Other Protected Areas

Peter Bridgewater
and Dan W Walton

Australian Nature Conservation Agency
GPO Box 636,
Canberra, ACT 2601
Australia

Abstract

Attention is drawn to the heterogeneity and complexity of the tropical savannas. Consideration is given to the role of national parks and other protected areas in such a landscape and what is meant by the term conservation. Various current and relevant issues are noted, including appropriate land use, reserve selection, land management objectives, limitations of protected areas and the importance of landscape rehabilitation. Aspects of ecologically sustainable development are noted and the involvement of the local community emphasised. The role of the Australian Nature Conservation Agency is considered.

1. SAVANNAS AS A LANDSCAPE

The tropical savannas are a complex heterogeneous landscape and their heterogeneity and complexity are not generally appreciated. Embedded within the tropical savanna landscape of Australia are highly diverse elements whose existence is derived from the interactions of the mix and the imposed cultural impacts over the past 50–60 thousand years. To the north, the savanna interacts with the tropical coastal fringe of Australia (similar landscapes occur in countries to the north) while to the south it links similarly with temperate landscapes. Perceptions of the tropical savanna are also complicated by ecological terms specifically relating to vegetation, such as grassland, woodland, etc. and land-use terms such as rangelands. In this paper, National Parks and other protected areas in the savannas will be discussed in the context of the biological resources of the whole savanna landscape and its use by people.

1.1 Appropriate use of lands and waters

More research and technology development is needed to ensure appropriate use of lands and waters. We do know enough now, however, to make more serious attempts to evaluate the efficiency and impact of existing water- and land-use practices. There is, however, no consensus on what is 'appropriate use'. If we are concerned about achieving a sustainable society, the definition of appropriate use will capture ecological sustainability.

The 'gamble' aspect of economic activity, the 'tyranny' of short term, narrow horizons and the inefficiencies and inadequacies of care, maintenance and use of natural resources must be minimised. Not only is accurate environmental information essential, but concise and meaningful analyses of the economic and social implications are essential. Where damage to the resource does occur through inappropriate use, an important issue is whether costs of rehabilitation should be passed on to the public purse, and if so, to what extent. The costs associated with prevention of inappropriate use will undoubtedly be less than the costs of rehabilitation. Although some view any inquiry into appropriate use as a potential threat to a variety of traditions and customs, there can be little doubt that a sustainable society must be, by definition, an equitable society (Lonergan 1994).

1.2 Present use of lands and waters

Present uses of the lands and waters of the savannas involve a few major industries (ASTEC 1993, Holmes and Mott 1993). There is no need to repeat observations on the impact of industry. Suffice to say that every human activity uses energy and produces waste: this is the essence of cultural shaping of the landscape. Nevertheless, the tropical savannas are not noted for agricultural output. The pastoral industry has a history and mystique, but its traditional form is being questioned (Hecht 1993, Milchunas and Lauenroth 1993, Morton 1993, Dodd 1994, Milton et al. 1994). Care and long term management of pastoral landscapes have traditionally been the roles of government agencies. Except for occasional significant downstream effects, mining activity is generally very site specific. The mining industry has become increasingly active in addressing the rehabilitation of mine sites. Tourism, or what is called ecotourism, is increasingly significant. Organised tours to established sites can be monitored for impact, but unstructured intrusion into the landscape can be expected to increase. Often overlooked in evaluations of present land- and water-use impacts are those associated with human settlement, including communications, energy supply and consumption, transportation and waste in its myriad of forms. National parks and other types of protected areas are tourist destinations and include human settlements.

1.3 Biological resources of the tropical savannas

In the tropical savannas, as in other parts of Australia, European settlers have concentrated upon those biological resources which they brought with them to Australia (cattle, sheep, cereal grains, vegetables, and many others). In pursuit of a niche in the world market place, biological resources of exclusive Australian origin have received very little attention. This is not too surprising, from both the historical and economic perspectives. There is now the need to look beyond the traditional and conventional, to new and different opportunities. If we are intent upon a sustainable society within an Australian milieu, the replacement of native plants and animals with alien species cannot continue apace nor can it be driven only by immediate economic gain in defiance of longer term consequences.

Making better use of native species or products from native species has potential for both increased economic value and sustainability of tropical savanna landscapes beyond National Parks and protected areas. However, the difficulty in developing new markets and achieving market penetration should not be underestimated.

1.4 Selection of Protected Areas

National parks and other types of protected areas have historically been established for a variety of reasons. The extent to which a park or other protected area incorporates the elements of the total variation of the tropical savannas is not only a function of physical geography, but of time and cultural influences. National parks and other types of protected areas seldom are managed to identical plans and each park or area is at one time the beneficiary and victim of funds allocated for management and the cultural impact on surrounding landscape elements. Similarly, the surrounding areas enjoy a reciprocal arrangement with parks and other types of protected areas. Within this

Table 1: Six IUCN (1994) categories of protected areas based on management regimes

I	Protected areas managed mainly for: (Ia) science or (Ib) wilderness protection
II	Protected areas managed mainly for ecosystem protection and recreation
III	Protected areas managed mainly for conservation of specific natural features
IV	Protected areas managed mainly for conservation through management intervention
V	Protected areas managed mainly for landscape/seascape conservation and recreation
VI	Protected areas managed mainly for sustainable use of natural ecosystems

highly complex landscape, therefore, national parks and other types of protected areas are not a single, simple part of the landscape mosaic, but are varied and variable elements in a constantly changing cultural landscape matrix.

At the present time, there is some confusion as to why new protected areas should be established. This confusion lies between those who see protected areas as cultural icons which must be enclosed to protect them from potential damage or destruction and those who view protected areas as essential components for the management and conservation of biodiversity in a landscape already 'managed' for tens of millennia.

The term conservation is often used carelessly. In the Global Diversity Strategy, the IUCN (International Union for Conservation of Nature and Natural Resources), the UNEP (United Nations Environment Programme), the WRI, in consultation with the FAO(Food and Agriculture Organization) and UNESCO, offer the following definition of conservation: *'The management of human use of the biosphere so that it may yield the greatest sustainable benefit to current generations while maintaining its potential to meet the needs and aspirations of future generations: Thus conservation is positive, embracing preservation, maintenance, sustainable utilization, restoration, and enhancement of the natural environment'* (Courrier 1992).

The IUCN has recognised the wide interests of society by recognising six categories of protected areas based on management regimes (Table 1). From the view point of conservation, however, high on the list of criteria there must also be emphasis on locking in the people of the savanna to the role of caring for their landscape.

1.5 Limitations of protected areas

Change is an inherent quality of living systems. Too often protected areas are regarded as static displays and, therefore, all change is viewed as undesirable. This naive view contributes to false expectations of security and unwarranted fears of change. Protected areas interact with the surrounding landscape matrix and the activities of people generate direct and indirect impacts. All of Australia's landscapes are cultural landscapes and, therefore, require active management. We must be careful not to raise the false hope that biodiversity can be 'preserved' by a network of small 'protected areas' which are assumed to be immune to all that surrounds them. Arguments about the importance of size of protected areas and the importance of specific site species richness must not be allowed to obscure the importance of the total landscape and the fact that all plant and animal associations exist in a broader framework of time and space.

Human population levels can be expected to increase. Fragmentation of the landscape will increase, new landscape connections will be established and other forms of cultural impact will increase. Protected areas unaligned with broad area planning will be of doubtful value as repositories of biodiversity, nor can they be expected to have a significant conservation role. The tropical savannas of Australia are not distributed according to political jurisdictions or land tenure and all elements of the landscape matrix, including protected areas, are vulnerable to differing and uncoordinated political decisions emanating from various tiers of government. Political boundaries are virtually irrelevant to gene flow, species movement, community structure or human impact.

2. DIFFERING CULTURAL LAND MANAGEMENT OBJECTIVES

As Walton (1994) notes, there are two major cultural views towards the landscape operating throughout Australia. There are a variety of options as to how one may view this situation (Nutting 1994), but the challenge is to find the complementary aspects of each that contribute to a sustainable society. The tendency all too frequently is to dwell upon areas of divergence as potential conflicts. We must recall that not only are the landscapes of Australia anthropogenic, but also different historical management regimes have and do contribute to the observed biodiversity. Biodiversity includes harvested species, whether kangaroos or domestic cattle, native yams or introduced sweet potatoes. Whether the issue is the harvest of traditional native foods or harvest of introduced crops, careful consideration must be given to the sustainability of the practice. The partner of rights is responsibility. The major shared cultural land use objective is ecological sustainability.

2.1 Landscape rehabilitation and off-reserve conservation programs

As noted above, the mining industry is making a significant contribution to both the research into landscape

rehabilitation and the actual process of rehabilitation. While the traditional approach of 'green' organisations has focused on parks and other types of protected areas, an approach to which Government has responded, Government has also responded to approaches from the rural industries by the development of significant off-reserve programs. Not surprisingly, there has been an almost exclusive focus on the 'selection' of relatively undisturbed species-rich areas for the creation of protected areas. Insufficient attention has been devoted to 'protecting' those areas undergoing rehabilitation, unless the intention is to simply *'let nature takes its course'*. New combinations and new associations of species will emerge in the rehabilitation process and their vulnerability is a matter of concern.

Similarly, sound conservation practices should not be confined to some mythological pristine place, but we must anticipate that significant elements of the flora, fauna, lands and waters of Australia will be in association with urban, suburban and agricultural areas. Rehabilitation will be an on-going activity, one in which more industries must follow the creative lead of the mining industry (Geiser 1991, Kleiner 1991). The responsibility for the house-keeping cannot be left solely to governmental agencies. Australia's environmental quality cannot be reduced to some governmentally owned minor percentage of the total area. All of Australia is important and all Australians have a role to play (Mott and Bridgewater 1992, Walton 1994).

3. ECOLOGICALLY SUSTAINABLE DEVELOPMENT (ESD) IN CULTURAL LANDSCAPES

The landscapes of Australia are anthropogenic and, as such, require management. Those ostensibly concerned with biodiversity focus on areas where human impact appears less severe, but large sections of Australia have been heavily modified for agricultural or other human settlement purposes. At the present time, considerable focus is on species which are not native to Australia. Many of these species have intentionally been introduced as food, fodder, fibre, ornamental plants and pet animals. Unplanned introductions include parasites and peri-domestics (species or populations of species which live 'about people' and includes commensals or those which thrive on human habitation and disturbance, but cannot be viewed as 'domesticated') or free-riders. Whatever their origin and however their entry, many are regarded as pests (Lonsdale 1994). Without doubt, many of these introduced species rely upon environmental disturbance of human origin for their existence or persistence while others are efficient invaders (Ghersa *et al.* 1994). New associations and combinations have become established and most are poorly understood. Very little work has addressed their potential value in conservation such as those ideas canvassed by Crosby (1986). There are no well established and widely accepted criteria for deter-

mining whether control could or should be applied, and if so, for assessing the type or extent of that control (Bridgewater *et al.* 1992). Any model for the ecological sustainability of Australia must be firmly based on the premise that all of Australia is anthropogenic, that no part or portion is structurally or functionally independent of adjacent parts or portions and that active management is as necessary to ensure the quality of the soil as to secure the fate of wildlife.

3.1 Ecologically sustainable development in changing landscapes

There is no doubt that some native Australian species are becoming peri-domestics while others also have become pests (Garrott *et al.* 1993, Lonsdale 1994). Increased fragmentation, isolation and new connections are resulting from current land and water use practices such as clearing for agricultural, pastoral or housing purposes. While immediate financial or other cultural gain is not without merit, the consequences of rapid and irreversible change must be carefully evaluated. The 'green' movement is fond of reminding us that extinction is forever, while some scientists titillate us with the promise of genetic resurrection à la 'Jurassic Park'. Forever is a long time and resurrection, if possible, could prove to be prohibitively costly. If we are wise, we will work with change by retaining natural options, i.e. retain biodiversity. Among the changes will be alterations to natural cycles, trends and chaotic events. These must be understood and distinguishable from induced deleterious changes of human origin (for some of the problems in dealing with a cycle of relevance to the tropical savannas, see Kerr 1993). Human activity will exert increasing pressures on evolutionary mechanisms, but there is no reason to assume that all these pressures are evolutionarily deleterious i.e. lead to extinction of species (van der Maarel 1993).

3.2 Ecologically sustainable development, Parks and other protected areas

The early 1970s saw the emergence from the UNESCO of the biosphere reserve model, the first internationally proposed attempt at integrated planning for ecological sustainability. That the model, perhaps ahead of its time, languished in relative obscurity for a few years, was to be expected, but the concept has experienced, with some refurbishment, recent rejuvenation (Ishwaran 1992). The model is a landscape management plan which clearly integrates the role of protected areas within the landscape mosaic. Prime Minister Keating's *Statement on the Environment* delivered at Adelaide in December, 1993 announced a major ESD bioregional initiative involving the acquisition of Calperum Station. Similar local community, State, Federal, NGO (Non-Government Organization) and international partnerships may offer significant opportunities unattainable by more solitary efforts. Regional perspectives of Australia such as that by Thackway

(1992) and Thackway and Cresswell (1994) provide information critical not only to the establishment of a national system of protected areas, but for management of biocultural regions for Ecologically Sustainable Development (Bridgewater et al. 1992). The traditional roles of parks and other types of protected areas remains valid, but fresh appraisals in a long term ESD perspective are warranted.

4. RESEARCH AND TECHNOLOGY NEEDS FOR MANAGEMENT OF PROTECTED AREAS

Two significant points relating to the research and technology needs of protected area managers emerged in the ASTEC (1993) inquiry:
(a) the information needs of managers as perceived by research organisations were not realistic and research tended to be aligned with interests of the researchers or their corporate culture;
(b) research results are most often presented in a form most beneficial to other researchers rather than in an information form readily useable by management.

Management is by necessity concerned with the social application of information derived from research. Technology comprises the use of various tools for solving problems and is not without costs to the environment (Cramer and Zegveld 1991). Research and technology are part of the armory of management. No doubt, researchers view the information needs of managers as not addressing 'basic' or 'fundamental' scientific issues. This is hardly a surprising conclusion since management primarily is concerned with crises involving the immediate wants and needs of people. Researchers and their parent agencies should have the ability to put the information needs of management into a 'basic' or 'fundamental' science framework while delivering readily useable information to management.

4.1 Setting R and D priorities

Scientists have traditionally regarded themselves as best equipped and placed to determine priorities (Aitkin 1994), but there is no evidence that scientists, or bureaucrats for that matter, are ideally placed to determine priorities. The citizenry generally are inadequately informed to make deep and meaningful judgements. They do, however, pay for a very large proportion of the research and appropriate consultative processes should be established (Miller 1994). The biocultural region comprises a framework for participation of the local people, not only in establishing priorities for research, but in the use of the products of research and technology development. Research and Development strategies for the complex tropical savanna landscape, as for all other major landscapes, must be part of a national strategy which reflects national goals (Aitken 1992, 1994, Bridgewater et al. 1994, Walton 1994).

As we and others (ASTEC 1993) note, the tropical savannas are not a homogeneous landscape. In addition to political constituencies, there are major ecological complexes such as the coastal zone, freshwater drainage systems, escarpments, rainforest patches, open grasslands and various woodland patterns. Industries and communities most certainly have interests. All will have advocates and priorities. Priorities set in a framework of long term national goals can reduce the chance of politicised focus on short-term crisis-of-the-week events (Abelson 1994) or knee-jerk responses to special pleading by narrowly focused interest groups.

4.2 Monitoring and long-term ecological research

Whether gathered or acquired through experimentation, all data are not equal. Priority must be given to the determination of what environmental data will comprise a baseline dataset. Such a data set should facilitate the detection of change. Data collected over time and space are essential if we are to detect change and attempt to attribute or determine the causes of change. At the present time, we have no idea of the status of most ecosystems, species or genetic systems. Standards are required for the collection, formatting, storage and care of environmental and biological data sets. Many of these critical issues are addressed by Busby and Walton (1994) and Miller (1994). There is no need to repeat these reviews here, but suffice to say that agreement on a basic dataset, the establishment of standards, custodianship and public access are central questions.

Long-term ecological research has a significant contribution to make to the total data, but careful consideration must be given to not only the kinds of long-term ecological projects which will contribute to our understanding of environmental changes, but to the priorities accorded to such projects. While the 'scientific merit' of the projects should be beyond reproach, the projects must contribute to the national goals. What must be borne in mind is that the rate of anthropogenic effects on the landscape may exceed any expectations allowed in the experimental design — or monitoring program (Bastian and Bernhardt 1993).

4.3 Infrastructure

The total land mass and marine areas which comprise Australia and her interests are huge and the total population which serves as the revenue base for the costs of administering Australia and her interests is remarkably small. This is not the time or place to open the discussion on population, although the discussion is critical to the determination of national goals. Australia, however, cannot afford the luxury of great redundancy in bureaucracy, research or technology, nor can it afford the development of monolithic structures which replace national goals with corporate culture. The efficiency and effectiveness of infrastructure is absolutely crucial to the development of a sustainable society.

Duplication, diversion and retention of inappropriate symbols result in inter-governmental or inter-agency squabbles over jurisdiction and significantly increase the required revenue. We must start with a clear understanding of the ecological characteristics and capacity of landscapes such as the savannas.

5. REGIONAL ROLE OF THE SAVANNAS

Young and Solbrig (1993) provide a concise and comprehensive assessment of the world's savannas. For our consideration of the regional role of Australia's savanna, we have chosen to use their four final suggestions as reference points.

1. *'To begin with a careful assessment of the ecological capacity of savanna systems;'*

Studies such as that carried out by ASTEC (1993) make significant contributions to an accurate assessment of the savannas. Neither the savannas nor the available information and technology are static and periodic reassessment is required. Court decisions and legislation affect land tenure and cultural land use. Research, technology development and inquiries seldom take full cognisance of Aboriginal and Torres Strait Islander people's view or needs. Assessment is making value judgements and values are deeply rooted in culture. Productivity is measured against desired outcome. A long-term view is essential. A balance must be reached between national goals and the goals of the people who live in the savannas. Unrealistic expectations of the ecological capacity of the savannas will disappoint national achievement and alienate the people of the savannas.

2. *'To avoid subsidies;'*

In most instances, subsidies are used to put the best face on a less than ideal, generally unsustainable, situation. Unfortunately, the environmental impact of subsidies is poorly understood, but they do tend to defuse the need to gain an accurate picture of the dysfunction. Poor economic potential or viability may well be closely related to inappropriate land use or inappropriate infrastructure. Great effort must be directed towards attempts to capitalise on the 'natural' economic, social and environmental advantages of the savanna landscape.

3. *'To devolve responsibility for management to local communities;'*

One of the essential ingredients of the biosphere reserve model is the involvement of local communities in the management of the reserve. Central government agencies and local government are often reluctant to relinquish what is seen as 'control' to local communities, for such a step means a change of corporate culture and change induces anxiety. Nevertheless, local communities have been intimately involved in park management at Kakadu and Uluru National Parks. The local Riverland community, albeit at early stages of development, is actively involved in the ESD programs at Calperum Station in South Australia. South Australia is taking seriously the suggested role of the local community in the management of biosphere reserves.

Policies and programs which transgress local social customs and traditions are unlikely to succeed and may prejudice future efforts. Conservation and the accompanying facilitation of those social changes required to achieve a sustainable society must be positive experiences, focusing on the *do* instead of the *don't*. The savanna landscape is under the influence of episodic events and local communities, rather than some remote central authority, are best placed to deal with such events.

4. *'To provide these communities with opportunities to make a significant but sustainable contribution to their economy.'*

In a socio-economic environment seemingly divided between big business and big labour and where Government is the prize of the dominant of the day, the importance of local communities is often over-looked. We must ensure that the policies applied to the savannas (or any other region) do not encourage inappropriate land and water use practices and that information on sustainable practices is accurate and easily accessible. Although not often a popular course to pursue, up-grading local community infrastructure may free resources for other projects of value. Local communities are best placed to deal with local cultural and environmental constraints. The best role for government agencies may be as the provider of information, adviser or councillor. Seed money, financial management advice and market assistance may represent three important contributions from government towards development of sustainable communities. Local communities need to be aware of the consequences of human population growth (Bernard *et al.* 1989), especially given the characteristics of the savanna landscape. Growth at any price is highly unlikely to lead to sustainability. Among the many options for plans for conservation, protected areas and ecologically sustainable development, the biosphere reserve model incorporates not only a set of management goals, but options for achieving the goals (Walton *et al.* 1992, Longmore 1993).

6. CONTRIBUTION OF THE AUSTRALIAN NATURE CONSERVATION AGENCY

The Australian Nature Conservation Agency (ANCA) administers biodiversity programs that deal with nature conservation both within and outside protected areas and also takes responsibility for national roles in a variety of international conventions and agreements. The ANCA also liaises and cooperates with Commonwealth, State, Territory, and a broad spectrum of other public and private organisations,

agencies and companies with interests in nature conservation. There is little doubt that if the *National Parks and Wildlife Conservation Act 1975* was legislation of 1994, this Act would be termed the *National Biodiversity Act*.

The ANCA, in view of its responsibilities, tries to maintain a holistic view of nature conservation, while recognising that programs on particular issues such as feral pests, endangered species, marine mammals or bird migration serve to focus attention on nature conservation generally and on possible indicators of landscape health (see Ferguson (1994) for a discussion of some aspects of landscape health). Unfortunately, scientists and area managers have been slow to define those traits which can serve as measures of (and which can be monitored) landscape health. Understanding health measures such as morbidity and mortality within the landscape framework will be central to the development and implementation of appropriate land use practices and landscape rehabilitation. Supporting legislation will need to focus on ensuring the prevention of morbidity and mortality rather than the more conventional command and control approach, which often can do little more than administer after-the-fact punishment. To continue to demand ever more 'basic' research, while failing to act on what is already known, will not only fail to serve the national goal of ecological sustainability, but will increasingly marginalise science and scientists within Australian culture.

As Cortner and Moote (1994) point out, there are recent significant policy changes in the United States which reflect a paradigm shift away from the traditional one characterised by sustained yield. The same, hopefully, can be said for Australia. The tropical savannas of Australia cross internal political boundaries and are subjected to widely different cultural management practices. What is the future of the savannas? Whatever its future, people will make the decisions and be an integral part of its future (Grizzle 1994). Are we prepared to deal with new paradigms, are we as scientists willing to make the changes in traditional institutional and professional cultures to adequately face the issues?

The ANCA lies at the centre of these questions and its willingness to help set and facilitate the agenda for change is critical. Attempting to arrest change or to pretend that nothing has changed will doom any hope of ecological sustainability. The tropical savannas of Australia are a collection of very complex heterogeneous landscapes which occur in Queensland, the Northern Territory and Western Australia. Aboriginal and Torres Strait Islander and European land management practices and goals exist. Three major industries are dominant: mining, pastoralism and tourism. The opportunity exists for those who have and are shaping the future of the tropical savannas to develop a plan of management which, with periodic adjustments, will ensure the welfare of the savannas. The tropical savannas and the people of the savannas are part of the cultural landscapes of Australia. While local issues are important, the dominant forces acting on Australia and Australians are global, including economic development, international trade, population growth, environmental degradation and, quite probably, climate change (Houghton 1994). As goes the landscape health of the tropical savannas, so goes the economic, social and political health of all Australia.

References

Abelson, P.H. (1994). Science, technology, and Congress. *Science*, **263**: 1203.

Aitkin, D. (1992). Aligning research with national purpose. *Canberra Bulletin of Public Administration*, **68**: 110-115.

Aitkin, D. (1994). In search of a national research plan. *R and D Review*, February, pp. 25-26.

ASTEC (1993). *Research and Technology in Tropical Australia and Their Application to the Development of the Region. Final Report*. AGPS, Canberra.

Bastian, O. and Bernhardt, A. (1993). Anthropogenic landscape changes in Central Europe and the role of bioindication. *Landscape Ecology*, **8**: 139-151.

Bernard, F.E., Campbell, D.J. and Thom, D.J. (1989). Carrying capacity of the Eastern ecological gradient of Kenya. *National Geographic Research*, **5**: 399-421.

Bridgewater, P.B., Walton, D.W., Busby, J.R. and Reville, B.J. (1992) Theory and practice in framing a national system for conservation in Australia. In: *Biodiversity - Broadening the Debate: a trilogy of discussion papers*, pp. 3-16. Australian National Parks and Wildlife Service, Canberra.

Bridgewater, P.B., Walton, D.W. and Busby, J.R. (1995). Developing policy for managing biodiversity at landscape scale. In: *Biodiversity in Managed Landscapes: Theory and Practice*. Oxford University Press, (in press).

Busby, J.R. and Walton, D.W. (1994). A national biological survey for the United States? Comparable Australian activities at the national level. In: *Biodiversity - Broadening the Debate 3*, (ed Longmore, R.), pp. 4-11. Australian Nature Conservation Agency, Canberra.

Cortner, H.J. and Moote, M.A. (1994). Trends and issues in land and water resources management: setting the agenda for change. *Environmental Management*, **18**: 167-173.

Courrier, K. (ed) (1992). *Global Biodiversity Strategy. Guidelines for Action to Save, Study, and Use Earth's Biotic Wealth Sustainably and Equitably*. UNEP, Washington, D.C.

Cramer, J. and Zegfeld, W.C.L. (1991). The future role of technology in environmental management. *Futures*, **23**: 451-468.

Crosby, A.W. (1986). *Ecological Imperialism. The Biological Expansion of Europe, 900-1900*. Cambridge University Press, New York.

Dodd, J.L. (1994). Desertification and degradation in sub-Sahara Africa. *BioScience*, **44**: 28-34.

Ferguson, B.K. (1994). The concept of landscape health. *Journal of Environmental Management*, **40**: 129-137.

Garrott, R.A., White, P.J. and White, C.A.V. (1993). Overabundance: an issue for conservation biologists? *Conservation Biology*, **7**: 946-949.

Geiser, K. (1991). The greening of industry. Making the transition to a sustainable economy. *Technology Review*, **94**: 64-72.

Ghersa, C.M., Roushe, M.L., Radosevich, S.R. and Cordray, S.M. (1994). Coevolution of agroecosystems and weed management. *BioScience,* **44**: 85-94.

Grizzle, R.E. (1994). Environmentalism should include human ecological needs. *BioScience,* **44**: 263-268.

Hecht, S.B. (1993). The logic of livestock and deforestation in Amazonia. *BioScience,* **43**: 687-695.

Holmes, J.H. and Mott, J.J. (1993). Towards the diversified use of Australia's savannas. In: *The World's Savannas,* (eds Young, M.D. and Solbrig, O.T.), pp. 283-320. UNESCO, Paris and Parthenon, Carnforth, UK.

Houghton, R.A. (1994). The worldwide extent of land-use change. *BioScience,* **44**: 305-313.

Ishwaran, N. (1992) Biodiversity, protected areas and sustainable development. *Nature and Resources,* **28**: 18-25.

IUCN (International Union for Conservation of Nature and Natural Resources) (1994). *Guidelines for Protected Area Management Categories.* IUCN, Gland, Switzerland and Cambridge, UK.

Kerr, R.A. (1993). El Niño metamorphosis throws forecasters. *Science,* **262**: 656-657.

Kleiner, A. (1991). What does it mean to be green? *Harvard Business Review,* **64**: 38-47.

Lonergan, S.C. (1994). Impoverishment, population, and environmental degradation: the case for equity. *Environmental Conservation,* **20**: 328-334.

Longmore, R. (ed) (1993). *Biosphere Reserves in Australia: A Strategy for the Future.* Australian Nature Conservation Agency, Canberra.

Lonsdale, W.M. (1994). Inviting trouble: Introduced pasture species in northern Australia. *Australian Journal of Ecology,* **19**: 345-354.

Milchunas, D.G. and Lauenroth, W.K. (1993). Quantitative effects of grazing on vegetation and soils over a range of environments. *Ecological Monographs,* **63**: 327-366.

Miller, R. B. (1994). The role of information in public policy. In: *Biodiversity — Broadening the Debate 3,* (ed Longmore, R.), pp. 12-20. Australian Nature Conservation Agency, Canberra.

Milton, S.J., Dean, W.R.J., du Plessis, M.A. and Siegfried, W.R. (1994). A conceptual model of arid rangeland degradation. The escalating cost of declining productivity. *BioScience,* **44**: 70-76.

Morton, S.R. (1993). Changing conservation perceptions in the Australian rangelands. *The Rangeland Journal,* **15**: 145-153.

Mott, J.J. and Bridgewater, P.B. (1992). Biodiversity conservation and ecologically sustainable development. *Search,* **23**: 284-287.

Nutting, M. (1994). Competing interests or common ground. *Habitat Australia,* **22**: 28-37.

Thackway, R. (ed) (1992). *Environmental Regionalisation. Establishing a Systematic Basis for National and Regional Conservation Assessment and Planning.* Environmental Resources Information Network and Australian National Parks and Wildlife Service, Canberra.

Thackway, R. and Cresswell, I.D. (eds) (1994). Towards an interim biogeographic regionalisation for Australia: a framework for setting priorities in the National Reserves System Cooperative Program (Draft Offprint, Version 3.1 dated 11 April 1994). Proceedings of a technical meeting held in Adelaide at the South Australian Department of Environment and Natural Resources, 7-11 February 1994.

van der Maarel, E. (1993). Some remarks on disturbance and its relations to diversity and stability. *Journal of Vegetation Science,* **4**: 733-736.

Walton, D.W., Forbes, M.A. and Thackway, R.M. (1992). Biological diversity, environmental monitoring and the nature conservation estate in Australia: relationship to ecologically sustainable development. In: *Biodiversity — Broadening the Debate: a trilogy of discussion papers,* pp. 24-40. Australian National Parks and Wildlife Service, Canberra.

Walton, D.W. (1994). Australia, cultural landscapes and other Australians. In: *Biodiversity — Broadening the Debate 3,* (ed Longmore, R.), pp. 21-36. Australian Nature Conservation Agency, Canberra.

Young, M.D. and Solbrig, O.T. (1993). Providing an environmentally sustainable, economically profitable and socially equitable future for the world's savannas. In: *The World's Savannas,* (eds Young, M.D. and Solbrig, O.T.), pp. 321-344. UNESCO, Paris and Parthenon, Carnforth, UK.

Chapter 6

Tourism in the Tropical Savannas

Gerry Collins

*Undara Lava Lodge,
Mt. Surprise, QLD
Australia*

Abstract

Tourism is showing renewed growth in Australia generally but, more particularly, environmental and cultural tourism is showing spectacular signs of growth. The tropical savannas are well placed to cater for this expansion, particularly since large tracts of natural bush are still in pristine condition and evidence of aboriginal and colonial culture and heritage are well preserved and unexploited.

Although the area devoted to National Parks in the savanna region is being expanded, these traditional tourist destinations will not be able to cope with the greatly increased tourist numbers nor their demand for increased diversity of experiences. Residents and local communities in the savannas have the opportunity to diversify into ecotourism to take-up some of this demand for tourism, while in many cases being able to pursue traditional land use activities. An important question is how this form of ecotourism should be planned and structured. To date much of the debate on ecotourism has focused on accreditation of operators and policing of the industry to ensure sustainability and guard against over exploitation. The practical resolution is to charge the industry with quality assurance and sustainability.

A healthy, viable industry will ensure that investment in tourism infrastructure will follow which will result in better management of our resources in the long-term.

1. INTRODUCTION

There is no question that mankind has had a greater evolutionary impact on shaping the world environment over the ages than any other living species, and will continue to do so. Therefore, we must accept that the human race is, in itself, a major evolutionary force. Modern society demands that we manage this impact as best we can with the resources available, with the ultimate goal of sustainability.

I am not exactly sure of the definition of sustainability in this debate with reference to the evolutionary process. However, if we choose to consider sustainability in terms of two or three generations of Australians, and achieve our goals during this timespan, we will have achieved remarkable results when compared with the last three or four generations who, unfortunately, did not understand the ramifications of this impact.

Let me take you back on the evolutionary time scale to the arrival of Aborigines in Australia — something in excess of forty thousand years ago. It is probable that this was the first major human influence that shaped the face of the tropical savannas as we know them in Australia today. It was from this point on, that fire accelerated the evolution of the plant species which dominate our tropical savannas today. This continued with the commencement of white settlement less than one hundred and fifty years ago.

Settlement was preceded by intermittent visits from early navigators as far back as the early sixteen hundreds, as well as annual seasonal visits by Macassans also dating back to this time. Then followed more recently Cook's visits, and others, prior to colonisation.

The history of settlement has been well recorded and it is evident that in Australia's tropical savannas this was triggered by Ludwig Leichhardt's expedition in 1844 from the Darling Downs, terminating at Port Essington in December 1845. This was followed by the dramatic expedition of Burke and Wills in October 1860. Despite the failure of Burke and Wills, no other exploration had a more profound or important effect on the settlement of our tropical savannas, particularly in the north and west of Queensland. The various parties and expeditions that went in search of Burke and Wills were the impetus for much of the settlement in the savanna region that sees it populated today. Generally, settlement was based on the pastoral industry, interspersed with brief, intense periods of localised mining activity.

The environmental impact, from the days of early settlement to when mechanised technology became generally used in pastoral and farm management, was negligible compared with the impact since then. With the introduction and breeding of more environmentally adaptable breeds of stock and Government policies on closer settlement and reduced living area sizes there has been more intensive use of the land.

The challenge is to develop and introduce alternative land uses which are sustainable in the longer term, or greatly reduce the direct impact on the land through many of our present practices.

2. CURRENT TRENDS IN TOURISM IN THE SAVANNAS

Tourism in its cultural and heritage, environmental, sport and adventure forms surely must be an alternative land use worthy of consideration. For the descendants of the first inhabitants and the early European settlers, who have persisted, in both instances, through sheer tenacity, tourism must appear to be worthy of consideration for either alternative land use or diversification, or both.

Tourism is showing renewed growth in Australia. The Australian Tourist Commission has a current target of 6.5 million international arrivals in the year 2000 (this was prior to the announcement of the hosting of the 2000 Olympics in Sydney). If that original target is realised, and with the new markets such as Korea and Taiwan making this more of a possibility, tourism will be far and away Australia's biggest industry, contributing more than $29 billion to the national economy (Australian Tourist Commission 1991).

More specifically, environmental and cultural tourism is showing spectacular signs of growth — and our tropical savannas are well placed to cater for this expansion, particularly since large tracts of natural bush are still in relatively pristine condition and evidence of aboriginal and colonial culture and heritage is well preserved and unexploited.

Surveys and market segment studies, commissioned by the Australian Tourist Commission and other Government agencies, and conducted by reputable research organisations, investigating potential international and domestic market demand, have clearly identified increasing interest in outback, rural experience, sport and adventure, cultural and ecotourism which can all be found in our tropical savannas.

'Outback' is an image and location that the majority of Australians living in coastal and/or capital cities wish to visit and experience. Of a sample of 1450 Townsville and Cairns residents and visitors, 61.2% had 'visited the Outback' and all 'wished' to visit if the occasion arose (Black and Rutledge 1993).

The growing importance of the Outback as an Australian tourism product may be gauged from a number of perspectives. In the past years, many indicators have emerged regarding the importance of the 'Outback' to the tourism industry. These indicators are also very relevant to the tropical savannas:

- Queensland's Minister of Tourism announced on ABC radio (12/5/93) that tourism into outback Queensland has increased by 43% in the past year.

- Queensland Tourist and Travel Corporation (QTTC) adopted a very definite Outback regional focus in 1992/93 with Market Segmentation Studies for each region e.g. *Market Segmentation Study — Outback Queensland* (QTTC 1992a).
- QTTC (1992b) has produced *Matilda Highway*, a glossy, high profile book highlighting points of interest and tourist attractions along the Matilda Highway from Cunnamulla to Kurumba.
- Tourism South Australia (1993) has produced a similar guide book — *Visitor's Guide to Outback South Australia*.
- The NSW Tourism Commission produced similar volumes (1992a, 1992b, 1992c) for their rural/outback areas
- Australian Tourist Commission research (1993) has shown a strong preference for 'aboriginality' and the 'Outback experience' by the Asian market.
- The Commonwealth Department of Tourism in researching rural tourism for a discussion paper (1993) could find only limited information on what defined rural or outback tourism. They were forced eventually to draw an arbitrary boundary between coast/rural and rural/remote based on town sizes (>100 000).
- The Commonwealth Department of Tourism, as part of a $42 million Federal Government commitment to improving rural tourism, has recently engaged Australian Farm and Country Tourism Inc. (AFACT) to examine the needs of the Australian rural tourism industry. The consultancy is to be completed by October 31 and the consultancy findings to be incorporated into the National Rural Tourism Strategy due for release in December 1994.

While these potential domestic and international markets have been identified, we have only just 'scratched the surface' and, to date, have fallen well short in developing an adequate infrastructure to service and meet these needs. The demand is there, and strengthening, and we should ensure that an infrastructure is developed to provide the facilities and service to satisfy this demand.

There are real opportunities now to develop a community based industry in our tropical savannas which delivers a special experience based on this vast expanse of land with all its inherent biological and physical attractions and aboriginal and colonial history. It is a specialist market to conform with the tourism strategies and management programs of the local authorities and regional bodies of the various areas, based on the needs of the visitor and sustainable use.

All too often the history, culture and attractions of our savannas are taken for granted. We, as locals, often become complacent and fail to recognise, or overlook, the significance of our tourist assets, and neglect visitors to our region. Residents in our tropical savannas must be made aware, that, if they wish to benefit from the opportunities available from embracing tourism as an alternative to traditional methods of land use, they must learn to identify and list their tourist assets, recognise opportunities, initiate feasibility studies and support regional plans for tourism development.

2.1 Integrating tourism with current land use practices in the savannas and establishing appropriate structures

As we approach the 21st century, we must be acutely aware, and not ignore the demands, of a global society conscious of the shrinking natural assets of the planet. Many of these assets are shrinking because of basic human demands to meet the primitive needs of local inhabitants who, because of the state of their local economies and basic demands, have no hope of introducing environmentally sustainable management programs. In Australia, we do have an economy, and a standard of living which affords our community the opportunities to plan management strategies which will manage these assets in an evironmentally sustainable manner.

This new trend in tourism offers savanna-based communities opportunities to diversify into tourism, whilst continuing to pursue traditional and present land use activities during the transition. The potential new income streams immediately create new jobs and new opportunities for an improved local economy, whilst reducing the risk of over-exploitation of our natural resources. I believe we are at the crossroads, where some decision should be taken but first we must ask ourselves some basic questions:

How long will our present land use practices be sustainable and economically viable?

1. What are the alternatives available to us, or combinations of present and alternative land use practices?
2. Which sustainable land use practices will continue to provide our communities with a way of life and a standard of living to which they are entitled? How do we react to the political pressure to bring about these changes?
3. What is the future of tourism as a global industry?
4. Will tourism continue as a strong industry if the growth of the world economy slows or stagnates?

My concern is, because of economic pressures, costs of production, unreliable rainfall patterns or previously defined living areas proving non-viable, that in some localities already, some of our land use practices may not be sustainable.

Government policies to establish large areas of national park, specifically to protect bio-diversity, are often developed without proper consideration being given to the added pressure of increased visitor impact

and the very limited resources to cope with the management challenges on the ground. The international recognition of world heritage areas and wilderness areas of international significance increase visitor pressure even further and compound the problem. It is questionable whether government departments and agencies have sufficient resources to adequately meet these challenges as there will always be shortfalls in funding because of higher government priorities in other areas. The ability of local government and communities to fund the development of infrastructure to meet the increased demand to access areas not under government management programs is usually limited.

This issue brings into question the role of government and the amount of government intervention needed to better integrate the tourism industry into the savannas. Much has already been achieved in organising the tourism industry *via* a statutory structure through federal and state legislation overlaid by a network of voluntary industry organisations serving the needs of special industry groups, operators and regions. The present structure, with some fine tuning when necessary, should generally meet the needs of a growing industry. However, one of the impediments to co-ordinating an effective structure across Australia's tropical savannas is state boundaries. Joint state co-operation is needed in planning strategies for the management of this massive area of land. Host farm and farm stay operators and their organisations have a good network, but we need to go further and establish goals which specifically aim for sustainable management while catering for the needs of all types of tourism.

3. FORGING THE WAY AHEAD

My belief is that the solution is to design and develop a tourism structure which is community based, self-regulating, sensitive to industry needs, market driven yet protecting and managing the resource. The community base can be provided by land holders and members of the community being involved, for they are potential tourism operators, guides, and interpreters, hosts and service industry participants.

With the potential for economic rewards, free enterprise has the capacity to meet this challenge. However, it will not succeed without strategic plans and market studies, establishment of adequate facilities, networking and regional marketing, self-regulation and community co-operation.

I believe this type of tourism will be extremely sensitive to price competition. In fact, there is no place for discounting and profitless volume because of environmental impact and the fact that there needs to be profit margins in the industry to develop the regional infrastructure to service the market.

The Gulf Local Authorities Development Association (GLADA) has been instrumental in tackling these issues to bring about more integrated tourism to the savannas of the Gulf region of Queensland and the Northern Territory. In the mid-1980s GLADA realised, from previous observations of the development of tourism in other regions, that, inevitably, greater numbers of visitors would journey to the Gulf because of changing trends in the market.

Travellers, both domestic and international, with a new 'eco-awareness' were planning to visit the remoteness and natural beauty of the savannas. It was observed that visitors travelled to these destinations irrespective of the lack of tourist facilities and local infrastructure, and strategies to service and meet the needs of the visitor. Observations indicated that visitors were planning to holiday at some of these outback destinations in spite of the attitude of some locals who wished the visitor would go away and leave them alone! This 'head in the sand' attitude was not the answer so GLADA decided to be positive and plan for the inevitable.

As an integral part of developing tourism in the Gulf region, GLADA developed the Savanna Guide concept. The philosophy of the Savanna Guide system was to meet the needs of the visitor by providing a high quality interpretative guiding service, coupled with the responsibility of sustainable management of the attractions and the region.

The current mission statement of the organisation reads:
'*To be an economically sound, community based, identifiable, professional body maintaining high standards of:*

    Interpretation;

    Public education;

    Tourism resource management;

    Leadership and training

...and through the promotion of ecologically sustainable tourism principles to enhance and maintain the regional lifestyle and encourage the protection and conservation of the environment and cultural resources of the Gulf Savanna Region'.

The guide organisation was formed in 1988 by individuals embracing these ideals at high profile attractions at various locations across the gulf savannas. Savanna Guides Ltd. is the incorporated company and the commercial vehicle on which the organisation is structured. The company has its own constitution, rules and bylaws and has an annual general meeting at which a Chairman, Secretary and four additional directors are elected. The bylaws set out specific requirements for guide conduct, standards of interpretation, uniform requirements, training, promotion and marketing. This all sets the standards which allow the philosophies of the organisation to be put into practice.

An important part of the success of the Savanna Guide system is proper training and accreditation of Guides. All new Guides commence as trainees, irrespective of

previous guiding experience, except in the case where they may have worked at associated savanna guide stations. The trainee is then trained on the job for a minimum training period before accreditation as Savanna Guide Station Site Interpreter. This accreditation takes place when the resident Savanna Guide, or Guides, are confident that the trainee has attained a level of competency acceptable to the organisation.

New Site Interpreters are encouraged to apply for Inbound Tourism Organisation of Australia (ITOA) guide accreditation. It is the policy of Savanna Guides Ltd. to have all site interpreters and fully qualified Savanna Guides ITOA accredited. ITOA have a national guide accreditation system which is seeking to improve the standard of guiding service available in this country.

Once accredited as a site interpreter, the trainee is encouraged to attend Savanna Guide training schools which are held twice a year at different locations in the region. The schools are moved around the region to allow a process of guide station operational assessment to take place and to improve the general regional knowledge of all guides. A trainee who has Site Interpreter accreditation and who has attended at least one guide school a year for two years may be nominated for elevation to fully qualified Savanna Guide. The nomination is considered by the directors and an assessment of the applicants guiding competency is reviewed. If the application is successful, the particular guide is invited to become a Savanna Guide as defined in the Constitution and articles of association of Savanna Guides Ltd.

The organisation has two guide stations in the Northern Territory and four on the Queensland side of the border, and is actively encouraging other operators in the Gulf Savanna region to embrace the philosophies of the Savanna Guide organisation and open new guide stations. The aim of the organisation is to establish a network of guide stations and guides across the region to service the needs of visitors and provide an accurate and reliable interpretation service, as well as a system for safe passage of visitors through the region by referring them from guide station to guide station. This helps to improve the visitor experience, by access to local knowledge, local property attractions and opportunities which are not normally available to the average traveller through the region.

The guide service automatically provides a level of regional environmental management through supervised visitation, education of visitors in procedures of best and accepted practice and supervised visitor impact at sensitive locations.

The Savanna Guide system may not have all the answers for improving tourism in the GLADA region but it serves as a good practical example of how the gulf communities are planning for the inevitable flood of visitors to a region which could be typical of much of Australia's tropical savanna.

3.1 Developing a tourist attraction in the savannas — Undara Lava Lodge: a case study

The Undara Lava Tube is a natural geological feature of great attraction to tourists in the Gulf Savanna region. The development of a commercial tourist venture arose out of the need to organise the flow of people wishing to visit the Lava Tube, part of which was on our property, simply to control and reduce the impact to a significant, fragile, local attraction. The solution simply was to make commercial tours available and to integrate them with a management strategy.

I was impressed by the developing policies of GLADA and conveniently embraced many of their philosophies when planning 'The Undara Experience'. I used, as a foundation for the project plan, a report on a study conducted by the Pacific Asia Travel Association (PATA 1992) task force titled *Gulf Savanna Territory, Planning for Action in Tourism*.

The report was very practical and comprehensive, and made a total of 50 main and 14 subsidiary recommendations — and clearly identified most of the issues to be addressed as far as I was concerned. The next step was to approach 'The National Centre for Studies in Travel and Tourism' at James Cook University in Townsville and, with their assistance, produced *The Undara Experience Development Concept and Marketing Strategy*. This in turn led to an economic feasibility study and, finally, construction plans. During this process we had made two important decisions:

1. To use old Queensland Rail carriages and wagons as the major theme in our lodge construction;

2. To join the 'Savanna Guide' organisation and be part of the guide network that GLADA was establishing across the Gulf Savannah.

The decision to use rail carriages was influenced by the PATA report which clearly identified the need to preserve examples of architecture in the gulf region, particularly examples which reflected the heritage of the region. The fact that history records the construction of a railway in the gulf over 100 years ago and that rail development represented a quantum leap in transport in the Einasleigh/Mt. Surprise area in the first decade of this century was obviously significant.

The railway theme combined very comfortably with the lava tubes which resemble large railway tunnels.

Additionally, a marketing report identified our market as predominantly self drive and about the mid-range of the star rating, possibly two to three star. The rail car accommodation fitted very comfortably into this category, whilst the top end of the market (that is guests seeking four and five star accommodation) do not have a lot of trouble accepting this standard because of the

heritage value of the carriages, their low impact on the site and environmental compatibility.

We considered that site impact was important. After the site selection process was completed, it was decided that site preparation would be minimal. The construction plan ensured that tree removal was kept to a minimum and building sites were selected where space occurred naturally and then drawn on the site plan. The rail carriages being long and narrow could be easily positioned between the trees, further reducing the need to remove trees.

The carriages were very transportable which enabled much of the renovation, refurbishing and refitting to be carried out in a builders' yard, which reduced builders' debris on site dramatically. From the time the first carriage was positioned on site, the lodge was operational in six weeks and stage one was nearing practical completion, thus reducing site impact considerably when compared with a traditional building program.

4. Thoughts for the Future

The region is perilously short of resources, we lack a well defined tourism strategy, we do not have a developed, on-going, dialogue with neighbouring regions, we have a state border dividing the region, a large percentage of our Gulf community does not appreciate the potential regional economic benefit from a well organised tourism infrastructure, and so on. Many of these issues can be positively addressed in the longer term, but there are other issues which need to be addressed now, not only in the GLADA area, but generally in our outback areas.

Now is the time to take stock of present resources and tourist infrastructure and 'fine-tune' our present operations so that they function more efficiently and effectively. We need to get back to basics and develop our skills as hosts, guides and service providers in the real 'aussie' definition of 'country hospitality' to provide an identifiable, marketable, special tourism product which creates market demand. A worthy initiative in this area has been the introduction of the IOTA 'Aussie Host' training program. Similar schemes have worked well in Canada and New Zealand and have many positive attributes. The program is industry driven and accredited and is about adopting a culture which guarantees guest satisfaction. This program goes one step further than many of the training programs which are available today and is purely based on better customer service and a greater awareness of customer needs and expectations. This concept is one which could be further developed for inland tourism to ensure that tourism products available across our tropical savannas are quality controlled by the industry, and tailored to meet the needs of the market and managed to guarantee sustainability.

Much of the debate to date on ecotourism has been about accreditation of operators and policing of the industry to ensure sustainability and guard against over exploitation — surely legislative controls are not the answer. The practical resolution is to charge the industry with the responsibility of quality assurance and sustainability, and initiate the introduction of industry standards which will guarantee the standards which are monitored and reviewed by the industry with procedures to ensure that the necessary audits and assessments are carried out periodically with the result that good operators are recognised and rewarded through an accreditation process which is industry-endorsed and internationally recognised.

Society can then be confident that practical standards are in place, reinforced by local government licensing of operators, with industry endorsed quality assurance procedures in place. The market will reward the good operators — I am confident of this — as we have a new awareness developing where very discerning customers are demanding evidence of environmentally sustainable tourism practices.

A healthy, viable industry will ensure that investment in tourism infrastructure will result which will provide an opportunity for better management of our resources in the tropical savannas.

Acknowledgements

Many thanks to Neil Black for providing much of the tourist data in this paper.

References

Australian Tourist Commission (1991). *Annual Report*. Australian Tourist Commission, Sydney.

Australian Tourist Commission (1993). *International Tourism Marketing Manual: Marketing your tourism product overseas*. Australian Tourist Commission, Sydney.

Black, N. and Rutledge, J. (1993). *Outback Tourism: An Authentic Australian Adventure*. JCU Press, Townsville (in press).

Department of Tourism (1993). *Rural Tourism: Tourism Discussion Paper No 1*. AGPS, Canberra.

NSW Tourism Commission (1992a). *The Escape Manual*. NSW Tourism Commission, Sydney.

NSW Tourism Commission (1992b). *Farm Stays & Country Holidays*. NSW Tourism Commission, Sydney.

NSW Tourism Commission (1992c). *Outback, Bush & Adventures*. NSW Tourism Commission, Sydney.

Pacific Asia Travel Association (1988). *Gulf Savannah Territory, Planning for Action in Tourism*. PATA, Sydney.

Queensland Tourist and Travel Corporation (1992a). *Market Segmentation Study - Outback Queensland*. QTTC, Brisbane.

Queensland Tourist and Travel Corporation (1992b). *Matilda Highway: The Outback from Cunnamulla to Kurumba*. QTTC, Brisbane.

Tourism South Australia (1993). *Visitor's Guide to Outback Australia*. Tourism South Australia, Adelaide.

Chapter 7

The Australian Defence Force and the Future of Tropical Savannas

A. L. Barton
& Major J. N. McDonald

*Facilities Branch,
BASC, Lavarack Barracks,
Townsville, QLD 4810
Australia.*

Abstract

The Australian Defence Force (ADF) has had a significant presence in northern Australia since World War II. The current presence is a function of; the increase in population; the development of regional and urban centres such as Townsville and Darwin, increased utilisation of a number of primary resources and the implementation of the 1976 Defence White Paper.

The ADF manifests itself as a matrix of Defence facilities and infrastructure, equipment and most importantly Defence personnel and their families. The ADF is enhancing its presence with a shift in focus to northern Australia. This shift involves a permanent relocation of personnel, development of new infrastructure and an increase in training and exercise activity throughout the tropical savannas. At a glance, Defence owns and operates 52 facilities across the tropical savannas. These are staffed by personnel from 40 different units of the ADF. Large areas of Commonwealth owned or leased land (total 920 000 ha) are controlled by the ADF and designated for key training activities. Most ADF controlled land is the responsibility of the Army and is used for live-firing and associated training activities.

This paper provides a brief but comprehensive coverage of how Defence currently manifests itself strategically, socially and economically in the tropical savannas of northern Australia. The imperative for ADF to train and exercise in northern Australia is highlighted. Details are provided of ADF's approach to ecologically sustainable development. This is explored in terms of legislative compliance, ADF environmental policy, local Army land management policy and new research and technology. Case examples of economic impacts and environmental management activities are provided to give a finer focus to the broad issues discussed.

1. INTRODUCTION

Defence has a significant presence in the tropical savannas of northern Australia. That presence is dominated by permanent and temporary deployment of armed units of the Australian Defence Force (ADF). A brief overview is given of Defence, the ADF in particular. Discussion then focuses on two main aspects:

a) The ADF's role in defending Australia, including the communities, natural resources and infrastructure collocated in the tropical savannas of northern Australia. Discussion is in terms of personnel, facilities, the strategic policy driving that presence and what the future holds;

b) Defence as a landholder to accommodate ADF personnel, facilities and infrastructure and at a broader scale as land-user of the tropical savannas for training and exercise activities.

These aspects are detailed in terms of the present situation and future plans.

1.1 Australian Defence Organisation overview

The ADF is an Australian Government organisation responsible to the Minister for Defence. It's mission is: *"To promote the security of Australia, and to protect its people and its interests"* (Ayers and Beaumont 1993).

Defence is composed of a forces executive and three armed services, known collectively as the Australian Defence Force and closely interrelated is the Department of Defence. The Chief of the Defence Force and the Secretary of Defence have joint responsibility for administration, organised under 8 programs (Table 1). Each program plays a vital role in ensuring that the Defence mission is met. Defence has a staff of 91 858 and had an actual defence budget outlay of $9.2 Bn and $9.8 Bn in 1992–93 and 1993–94, respectively.

Table 1: Key responsibilities of the Chief of Defence Forces and the Secretary of Defence

Chief of Defence Forces	Secretary of Defence
Forces Executive	Strategy and Intelligence
Navy	Acquisition and Logistics
Army	Budget and Management
Air Force	Science and Technology

1.2 Defence Strategic Policy

The ADF is in a period of significant change, which can be traced back to recommendations provided from four pivotal documents:

- 1976 *Defence White Paper* (Anon. 1976)
- *Review of Australia's Defence* Capabilities (Dibb 1986)
- 1987 *Defence White paper* (Anon. 1987)
- 1993 *Strategic review* (Anon. 1993).

These documents have moulded policies that echo and enhance the lessons learnt during the defence of Darwin in 1941–42 and in later regional conflicts. Although overshadowed by the recent end of the Cold War, it is noteworthy that Australia's defence policy has been evolving independently for nearly twenty years, with a strong emphasis on regional stability. The cornerstone of Australia's defence is self-reliance within a framework of alliances and regional defence relationships. Australia's defence policy focuses on Australia's ability to defend itself with its own resources and to structure the ADF in terms of the perceived level of threat and the areas where that threat would manifest itself.

Because of our geographic location, conventional military attack against Australia would most likely be directed against the northern part of the mainland, its maritime approaches or off-shore territories ('the sea-air gap') and that it would come from or through the archipelago to Australia's north. Current analysis is not based on any perceived level of threat but a function of our location and the tactically sensible and practical passage for hostile arms against this continent (Ray 1993).

This policy is amplified through the process of 'policy transparency', which seeks to ensure neighbouring states understand Australia's policies in the region, and to feel assured that there is no covert purpose in Australian actions (Lewis Young 1994). An important component of ensuring 'policy transparency' is the involvement of military personnel from neighbouring nations in exercises, training, staff interchanges and shared information (Ray 1993, Lewis Young 1994).

To achieve self-reliance in defending the most likely area of hostility the ADF is in the process of shifting its centre of gravity, particularly ground-based components, to the north. The more significant activities associated with the shift to the north relate to changes in location and capabilities of Army and Air Force units as follows:

- The movement of the 2nd Cavalry Regiment to Darwin;
- Improved mobility of Army units based in the north;
- Relocation of a Brigade to the north by the end of the decade (including an armoured regiment with one regular tank squadron;
- Increase in major exercises undertaken in the savannas to better train and prepare the Army for land-based defence activities;
- An increase in the forward-deployment capacity of the Air Force, including the upgrade of RAAF (Royal Australian Air Force) Base, Tindal, Construction of RAAF Base, Scherger, near Weipa and RAAF Base, Curtin near Derby to complete the network of air bases across the savannas.

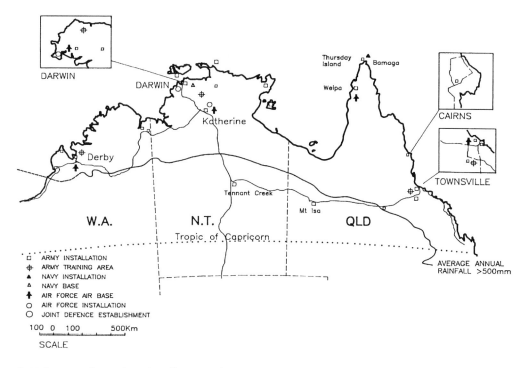

Figure 1. Major Australian Defence installations in the north.

ture Review (Ayers and Gration 1991a) and the 1992 *Defence Regional Support Review.* The *Force Structure Review* has several key elements, including the continuing shift of the ADF's focus to the north and west, enhancement of combat capabilities by redirection of resources from support areas, greater use of commercial infrastructure and the implementation of the Ready Reserve Scheme.

Implementation of the Defence Regional Support Review will improve regional administrative support arrangements. Its primary task was to consider the potential for joint service and integrated service/civilian administrative arrangements in the States and Territories and to develop a framework for the devolution of responsibility and accountability to the lowest practicable level (Podger 1993). A significant result is the establishment of Defence Centres in each of the States to replace former Military Districts and Civilian Regional Offices. In the case of the Army program this has meant that the Army's Logistic Command now administers Army through an Australia-wide regional network of ten Base Administrative Support Centres (BASC) and a Northern Region Administrative Support Centre (Grey 1993, Lewis Young 1994).

2. THE ADF'S CURRENT PRESENCE IN THE TROPICAL SAVANNAS

Defence has facilities right across the tropical savannas. The tropical savannas cover an area which can be broadly defined as that part of tropical Australia with rainfall greater than 500 mm, excepting the narrow strip of wet tropical coastline and rainforests. The main concentrations of the ADF are in Townsville, Darwin and Katherine. Other locations consist of surveillance facilities, air bases, administrative office accommodation for Army reserve units and Commonwealth land controlled by the Army or Air Force for Defence training activities. These aspects of the ADF's presence are presented in Figure 1.

In addition, there are a number of air bases located in northern Australia to meet the rapid strike and interdiction objectives of the Air Force. Tindal Air Base is the Air Forces main base in the north and is the hub of future northern operations supported by a number of bare bases located between Learmouth in Western Australia and Townsville in Queensland. The most recent development has been the construction of RAAF Base, Scherger near Weipa in March 1993.

The ADF presence corresponds with the main urban centres in the north. This is a function of the priority of defending key installations and the need for access to the accommodation, community and commercial support afforded by large urban centres. The following is a brief overview of the larger concentrations of the ADF's presence in the north.

2.1 Townsville

2.1.1 Army

The Army's presence in Townsville is dominated by the 3rd Brigade. It is the major combat component of

Defences 'Operational Deployment Force' and remains at the highest level of readiness for rapid deployment. There is also the 11th Brigade, an Army Reserve Formation with headquarters at Jezzine Barracks.

The combined personnel strength for the Army units in the Townsville area is currently 4109 (4034 Service, 75 civilians) or 12% of the total employed persons in the Townsville/Thuringowa region. Approximately 10% of the population of Townsville/Thuringowa is made up of Army personnel and their families. In addition there are approximately 1700 members of the Army Reserve that are drawn from the local community and therefore is a significant component of the social face of the Townsville area.

2.1.2 Air Force

The Air Force is represented in Townsville by units at RAAF (Royal Australian Air Force) Base Townsville, which is collocated at the Townsville City Airport. The base has a strategic importance and is "*an established air head for defence operations for the security of northern Australia and a stepping stone to RAAF base Scherger near Weipa*" (Warren 1993). The increased training imperative is demonstrated by the role RAAF Base Townsville plays in the continual process of RAAF training, exercises and refresher courses. There are three main units based at Townsville; 35 Squadron (Caribou Aircraft) with its pivotal role in the rapid deployment of combat elements of the 3rd Brigade, Number 1 Operational Support Unit and 323 Air Base Wing which has a similar function to the Base Administrative Support Centres of the Army. The RAAF directly employs 701 people (Warren 1993).

2.1.3 Navy

The Navy has no units permanently present in Townsville, but does use the Port of Townsville facilities during routine patrolling. The Navy's main location, within northern Queensland, is at HMAS CAIRNS situated in Cairns on the Wet Tropical Coast.

2.1.4 Housing

The Defence Housing Authority is responsible for providing off-base accommodation for defence personnel and their families. This involves the process of acquiring, maintaining and disposing of residential property. The Service Housing Authority performs a property management role (Warren 1993). The two organisations have a combined staff of approximately 180 people.

2.2 Darwin

July 1st 1993 marked a significant shift in emphasis to the north for operational command of northern based units. The Chief of Defence Forces assigned the commander of northern command (NORCOM) permanent army, navy and air force assets and a new administrative organisation based in Darwin (Lewis Young 1994). This force presently entails the following:

- Darwin-based 'Fremantle' class patrol boats;
- 2nd Cavalry Regiment armoured reconnaissance unit;
- NORFORCE, the Reservist Regional Force Surveillance Unit;
- 1st Combat Engineer Regiment, Darwin detachment;
- 7th Intelligence Company;
- 7th Military Geographic Information Unit;
- Air Transport Detachment;
- Northern Regional Administrative Support Centre.

These units undertake a range of land and sea based activities throughout the Northern Territory and the northern part of Western Australia. The Army is located at Larrakeyah and the newly constructed Waler Barracks built at a cost of A$64 million. The Air Force is located at RAAF Base Darwin and the Navy at HMAS Coonawarra and Larrakeyah. Northern Command's primary tasks include intelligence, surveillance, reconnaissance, and protective operations (Lewis Young 1994).

2.3 Regional presence in the tropical savannas

The ADF has a number of facilities in smaller regional centres, the majority of which are to cater for the needs of the Regional Force Surveillance Units. These units are an integral part of many local communities. They have operational responsibility for surveillance of the coast and hinterland of northern Australia, an area stretching from the Pilbara in Western Australia (Pilbara regiment), Darwin (NORFORCE) and to Cardwell in Queensland (51 Far North Queensland Regiment). These units are predominantly staffed by reservists with a wealth of local knowledge and skills in undertaking effective operations in the tropical savannas and the associated coastal zone. There are 500 reservists serving with NORFORCE and of these 110 are drawn from local Aboriginal communities. Their involvement in Defence activities also provides a valuable resource for educating other Defence personnel on survival and concealment in the demanding tropical savanna.

The surveillance units have a number of sub-unit Headquarters located in regional centres throughout northern Australia. Although logistically small the presence of these units is more obvious than other Army units because their activity in the savannas is continuous and relies on maintaining a co-operative relationship with other inhabitants, re-inforced through periodic surveillance operations and training activities. It is also a function of their greater integration with the civilian community.

In addition to the surveillance units there are telecommunication and electronic surveillance facilities that have small number of staff. The Army training areas are by far the largest properties.

3. ADF's LAND UTILISATION OF THE TROPICAL SAVANNA

The ADF uses both Commonwealth land for key training activities and accesses other areas of the tropical savanna for larger scale activities like the triennial Kangaroo exercise series. Of all Commonwealth organisations, the ADF controls the most area of Commonwealth land in Australia (ANAO 1992). The largest of these properties are controlled by the Army for the purpose of providing areas suitable for the continual training and exercising of ADF personnel. There are six Army training areas in the tropical savanna:
- Yampi (566 000 ha);
- Townsville (232 000 ha);
- Mt. Bundey (117 000 ha);
- Mt. Stuart (8143 ha);
- Kangaroo flats (5207 ha);
- Macrossan (388 ha).

In addition the Air Force has the Delamere Air Weapons Range (220 000 ha) near Tindal, Northern Territory.

These training areas are distributed across the tropical savannas (Figure 1) but usually in reasonable proximity to the urban centres that have a concentration of ADF personnel, particularly Darwin and Townsville. The training areas remote from concentrations of Defence personnel are utilised to a much lesser degree.

3.1 The Training Imperative

The continuous process of training and exercise is intended to maintain the degree of readiness needed to meet the three predicted levels of potential conflict: low-level conflict, escalated low-level conflict and to a lesser extent more substantial conflict (Anon. 1991). The ADF also requires continued access to much of the savannas to host the larger exercises, for example the Kangaroo series of exercises. Because the objectives of such exercises cannot be met within the confines of Defence owned property, this discussion will cover the issues of environmental management of both ADF owned land and other areas utilised for training purposes from the perspective of sustainable land use for training and exercise activities.

There are two types of training:
a) individual training to develop and maintain individual skills, particularly in the use of live firing of small arms, and collective training which aims to ensure that individuals can operate cohesively in groups, using a larger range of weapons and transport methods;

b) Collective training involves the live firing of individual weapons, mortars, artillery and tank guns; the detonation of live demolition, the construction and demolition of defence emplacements; and the tactical manoeuvre of personnel and the co-ordinated manoeuvres of tracked or wheeled vehicles, aircraft and personnel.

3.2 The Training Cycle

The training cycle of the 3rd Brigade is an example of the continuous training the Army needs to undertake to ensure it can always meet the objective of preparing a defence force capable of undertaking a land-based defence of Australia and its territories.

Each calendar year is divided as follows for the purposes of training:
a) Feb–May: section and platoon-level training (9–30 personnel);
b) May–June: sub-unit (company, battery, squadron) level training (100 personnel);
c) Jul–Aug: unit (regiment and battalion) training (350–600 personnel);
d) Sep–Oct: brigade and higher level exercises (3000 personnel and upwards);
e) Nov: specialist courses for individual soldiers;
f) Dec–Jan: leave.

4. SUSTAINABLE DEFENCE USE OF THE TROPICAL SAVANNAS

Training and exercise activities have the potential to cause environmental impact, particularly the impact of high explosives, off-road movement of tracked vehicles, construction and maintenance of facilities including artillery firing points, aircraft landing strips, accommodation and access roads and tracks. However, there is a strategic military benefit in minimising impact of military activity because of the need to conceal from a number of forms of surveillance, including airborne and satellite remote sensing devices. This aspect of minimising impact is drilled into all personnel of the ADF as it will influence their survival in times of conflict.

In addition to the strategic necessity for minimal impact, the Army applies a set of environmental management principles to further minimise the potential impact of training and exercise activities. These principles form the framework that Defence uses to meet its commitment to ecological sustainable development. The framework is structured to reflect policy commitments at a number of levels within the Commonwealth Government, Defence as a whole and at the local level. For example, the Army unit responsible for managing military training areas in North Queensland. The framework includes the following.:
a. *Commonwealth Policy Framework*
- *Environment Protection Act, Australian Heritage Commission Act;*

- National Strategies for Ecologically Sustainable Development, Biodiversity, Waste Minimisation and the Decade of Landcare.

b. *Defence Policy Framework*
- Defence Instructions on the Environment, Heritage and the management of same.

c. *Local Army policy*
- Range Standing Orders'
- Environmental Management Plans.

4.1 Environmental management by Defence as a whole

A framework for environmental compliance is provided by Defence Instruction Environment and Heritage Protection (DI (G) Admin. 40-1) (Ayers and Gration 1991b). This document details the procedures for environmental impact assessment and clearance of Defence activities and proposals to ensure that Defence's statutory obligations for environment and heritage are met. The principal Commonwealth legislation is the *Environment Protection (Impact of Proposals) Act* 1974 (EPIP Act) and the *Australian Heritage Commission Act 1975* (AHC Act). Defence also endeavours to comply with relevant state legislation and local government ordinances.

In addition there is a Defence Instruction being prepared that details the process for preparation of Environmental Impact Assessment (EIA) reports and development of Environmental Management Plans (EMP) that focus on the issues identified in the EIA. These reports are currently being developed for a number of larger Defence controlled properties including Army training areas. An EIS and EMP has been completed recently for Mt. Bundy Training Area in the Northern Territory. Another prime example of the EIA/EMP is the current proposal for Townsville Field Training Area (TFTA) which is detailed further in this section.

Defence is currently in the early stages of developing an environmental management information system (EMIS) to assist training area staff in meeting environmental management objectives. The final product, Defence Training Area Management Information System (DTAMIS), will assist in meeting those objectives. There are several systems under trial and these are providing valuable supporting information required in setting the parameters for the final configuration of DTAMIS. The system will be used to record bookings, danger templates, weed infestation, feral animal impacts, wildfires and a number of other sets of spatially referenced data which will be stored, accessed, analysed and updated on a Geographic Information System (GIS).

In addition CSIRO has been actively involved in land management at Puckapunyal Range in Victoria. Key projects include a land management monitoring program and a land management advice system (LMAS) which uses artificial intelligence to predict the effect of armoured units (Tanks and Armoured Personnel Carriers) on the soil surface under a range of soil moisture conditions. These projects will be modified for application on training areas in the tropical savannas.

4.2 Environmental management for Army Training Areas

The day-to-day management of Army training areas is the responsibility of personnel within the local Base Administrative Support Centre (BASC). The range control officer and his staff have responsibility for bookings, road maintenance, security, environmental management tasks, etc. There are differences in the number of staff assigned to a particular training area. This is a function of the variation in types of training activities undertaken in a particular training area and the amount of use.

Local policy for environmental management in Army training areas is driven by compliance with the Defence Instructions detailed above and through the promulgation of a locally prepared document for each training area known as Range Standing Orders (RSO). RSO's are written for a specific training area and designed to address a comprehensive range of issues including bookings, communications, environmental compliance, un-exploded ordnance, safety and access. Each unit that needs to utilise a training area must be fully conversant with RSO's and commit themselves to compliance with its contents.

With the development of EMP's for each training area, RSO's will be adjusted to meet any changes detailed in the EMP. The two documents will then work in tandem to ensure that all aspects of range use and management are appropriately addressed.

4.2.1 Case Example: North Queensland Training Areas (NQTA)

The training areas within the area of operations of the Base Administrative Support Centre-Lavarack (BASC-L), Townsville are: Townsville Field Training Area; Mt. Stuart Training Area; Tully Training Area; Cowley Beach Training Area, and Macrossan Training Area. Tully and Cowley Beach are within the wet tropics and the remainder are within the tropical savannas of north Queensland. These training areas are directly managed by North Queensland Training Areas (NQTA), a sub-unit of BASC-L and undertake a range of duties related to administration, security, safety, land management and repairs and maintenance.

The process of managing each area begins with the promulgation of Range Standing Orders for each training area to all units that intend to train in North Queensland. These documents set the basis on which

units will be permitted to undertake training activities in terms of responsibilities, general restrictions, forward booking, environmental compliance and management, range control procedures, safety, communications, storage and safety of ammunition, fire orders, medical and hygiene. Each unit has to comply with a unit/activity specific environmental compliance certificate and must appoint a environmental damage control officer. RSO's detail the key issues relating to conservation of the particular training area including such issues as waste disposal, digging, fauna and flora protection, vehicle movement, prohibition to key sites of heritage significance and minimisation of weed entry and removal.

Before the exercise, the unit is briefed to ensure that RSO's are clearly understood, any confusion can be redressed and such issues as fire hazard, which vary seasonally and in some instances daily, can be discussed. Once an exercise is completed a range inspection is undertaken and clearance may then be given by NQTA staff.

This comprehensive process is designed to avoid most forms of impact before they occur. However, there are still training activities that by their very nature incur a degree of impact. Particularly off-road activities of tracked vehicles and to a lesser extent live-firing involving artillery, vehicle mounted guns and aerial bombing. These activities are managed by dispersal, rotation of areas and by restricting access during the 'wet season' (December–March).

There are instances where certain activities are concentrated at key locations. The aim is to contain impact so that the remainder of the training area is better conserved. An example is that most forward observer training (artillery support) is mainly undertaken at one location in TFTA. This reduces the amount of impact area which is in view, therefore concentrating Artillery and Air Force activity to a smaller part of the impact sector. These locations are referred to as 'sacrifice areas'.

The issue of un-exploded ordnance (UXO) is a difficult problem to resolve. By law, live-firing can only be take place on Commonwealth land. The problem occurs when a training area is no longer required by Defence and is considered for disposal. Contamination by UXO makes sale of land and its suitability for alternative uses difficult. Impact areas are kept as small as possible to concentrate incidences of UXO and detailed records are kept of their location.

Environmental management of each training area is being further enhanced with the progressive development of Environmental Impact Assessments (EIA) and preparation of Environmental Management Plans (EMP) that address the issues raised in the EIA. The EMP's will be implemented by Environmental Officers specifically appointed for this role.

4.3 External auditing of management performance

An audit was conducted by the Australian National Audit Office (ANAO) in 1991–92 on the management of Army training areas (ANAO 1992). This document provides a basis that the Army is using to enhance its management of training areas. The ANAO recognised that the Army training areas were generally well managed at the local level, although more control was needed at the corporate level. In addition, concern was raised over the pollution and land contamination impact due to un-exploded ordnance (UXO). A recommendation was made that high explosive impact areas be kept as small as practicable. Most significantly the ANAO recognised the need for implementation of environmental management information systems (EMIS) for each training area to enhance site monitoring and management.

The Army is in the process of implementing 40 of the 43 recommendations contained in the ANAO report. The DTAMIS project, already detailed, is one of the Army's activities identified in the ANAO recommendations. Another is the recent appointment of an Environmental Officer or Land Manager, attached to each BASC unit, who has responsibility for the implementation of EMP's for training areas in a defined area.

The implementation of an EMP involves a routine (3 years) inspection by an external auditor to ensure that the EMP is being implemented and that the EMP is appropriately written to provide policy guidance in addressing relevant land management issues. In addition there is the initiation of Community Advisory Committees which will have input into the environmental management of training areas in a particular area. This will provide a better flow of information in both directions so that the community is aware of the management achievements of range control staff and the community are involved in the process of enhancing management activities and priorities.

4.3.1 Case Example: Townsville Field Training Area (TFTA)

The Townsville Field Training Area (TFTA) is a proposed expansion of an existing facility, High Range Training Area (HRTA). The objective is to disperse the impact of activities currently concentrated on HRTA and to reflect the increase in training and exercises flowing from the ADF's policy on increasing its presence in the north. The following is a case example of the acquisition and environmental impact assessment process that has been undertaken to date.

TFTA comprises the pre-existing HRTA (48 683 ha), Dotswood Station (155 907 ha) and a Special Lease over part of Bluewater State forest. The Special Lease is wholly within the *Wet Tropical Rainforests of North Queensland: Interim Register National Estate 'Listed Area'* and partially (11 600 ha) within the *Wet Tropics of Queensland' World heritage Area* (WTQWHA) as

The Future of Tropical Savannas

A draft EIS was prepared and made available for public review in late 1993. The document included a comprehensive description of Defence's proposed use of TFTA and a detailed environmental resource inventory was undertaken covering both the biophysical and socio-economic environments. Preparation of the document involved a community consultation programme to raise community awareness of the proposal and obtain public comments on the proposal for incorporation into both project planning and the decision making process (Dames and Moore 1993).

From matching proposed activities with the detailed resource inventory a land capability map for each of the land types within TFTA was prepared. This forms the basis for defining the sustainable development of the resources. A number of issues were identified that will need to be addressed in the Environmental Management Plan to meet the objectives of ecologically sustainable development.

The more significant findings of the draft EIS included the following:

a) The identification of a number of previously unknown archaeological sites of interest to the local Aboriginal Community. The three most important sites were on the existing HRTA and were found to be well protected under existing Range Standing Orders which prohibit live firing over or near the sites and forbid access to the sites by unauthorised personnel. The sites are accessible to representative members of the local Aboriginal communities in Townsville and Charters Towers.

b) There were also sites of heritage significance due to their role in the early phases of grazing and mining industries in the area. This includes, Dotswood Homestead, the remains of two townships which were part of the coach road network of the last century and the Argentine Chimney which was part of a facility used for extracting silver from crushed ore.

c) The soils of TFTA are particularly susceptible to erosion if surface cover is reduced significantly. This is exacerbated by the steep slopes along the eastern part of TFTA, particularly within HRTA.

d) The vegetation consists of a number of open eucalypt woodland communities. Of lesser extent, but equal importance, are areas of semi-evergreen vine thickets, rainforests, tall open forests and threading through the whole of TFTA, riverine forests. In the EIS, vegetation was assessed in terms of its regional significance, conservation status (National, Regional, Local) and scientific value.

e) Locally rare plant species identified were *Eucalyptus moluccana* and *Acacia nesophila*. Two other species listed in Schedule 1 of the *Endangered Species Protection Act, 1992* are predicted to occur in TFTA, but their presence is

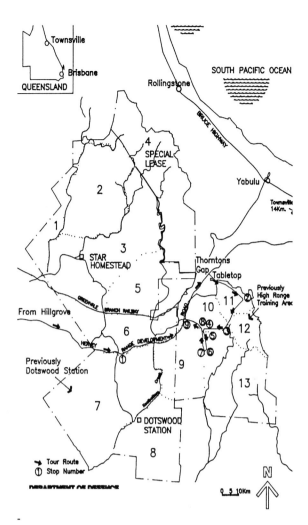

Figure 2. Department of Defence Townsville Field Training Area

shown in Figure 2. HRTA has been used by Defence since 1965; it is the manoeuvre area for the Townsville based 3rd Brigade and other units based throughout Australia. It is used for infantry (up to battalion strength), cavalry (up to squadron strength), artillery (up to regiment strength) and field engineering activities (Dames and Moore 1993). It is a particularly important training area as the 3rd Brigade maintains a higher level of readiness than any other Army unit. HRTA is not large enough to cater for Brigade level exercises on a sustainable basis (Cooksey 1988). During 1988 Dotswood Station, adjacent to HRTA, became available for purchase.

During the process of acquisition and establishment of the enlarged training area, Defence has complied with the *EPIP Act 1974* discussed in earlier sections. The Commonwealth Environmental Protection Agency directed Defence to prepare an Environmental Impact Statement (EIS) in relation to the proposal to utilise TFTA for training purposes.

yet to be confirmed. A number of declared and environmental weeds were also identified and their impact on the environment highlighted.

f) 295 species of vertebrates, from 83 families (excluding fish), were recorded in the fauna survey. The list includes a number of species which warrant special attention due to their restricted extent or reliance on a particular habitat. The key habitats that require effective management to ensure population stability are the Rainforest and associated Tall Forests, as well as the widely distributed small patches of semi-evergreen thicket and riverine forest. Feral animals were recorded and their impact on the biophysical environment of TFTA and surrounding neighbours highlighted.

g) Other issues covered include water quality, fire, acoustics and transport.

The key concerns raised in public submissions were conservation and protection of a number of species of local significance or listed in schedule 1 of the *Endangered Species Protection Act 1992*, restriction of access, protection of a number of sites of interest to the local Aboriginal communities, the future of grazing, waste/pollution, pests, fire management, water quality and importantly, community consultation.

A supplement to the draft EIS is close to completion and incorporates comprehensive responses to the submissions provided during the public review period. In addition, the supplement incorporates a draft Environmental Management Plan (EMP). The supplement will be referred to CEPA and the two documents (draft EIS and supplement) will form the final EIS. CEPA then examines the supplement to meet the following criteria:

a) to ensure the objectives of the Act have been met with respect to the proposal;
b) to determine whether additional environmental information is required;
c) to prepare an Environmental Assessment Report;
d) to formulate any recommendations or suggestions on the environmental aspects of the proposal which may be applied in association with the approval of the proposal.

A Community Advisory Committee will be established in line with the approval of the EIS and preparation of a final EMP for TFTA.

4.4 The ADF exercising in the tropical savannas

As detailed in 3.1 there are large scale exercises that require access to land outside the boundaries of Commonwealth Land. This revolves around two specific types of activities, the large scale brigade level, or larger, exercises culminating in the triennial Kangaroo exercise series and the much smaller unit, sub-unit, platoon and section training regularly undertaken by the Regional Force Surveillance Unit's as part of their surveillance role.

4.4.1 Case Example: Kangaroo Exercise Series

A notice of intention (NOI) was submitted to the Commonwealth Environmental Protection Agency in 1993 which detailed the salient components of a proposal by Defence to undertake an Australian Defence Force joint exercise in the northern areas of Western Australia, Northern Territory and the Cape York Peninsula during July/August 1995. This was preceded by a period of close consultation with representatives of forty-seven Federal, Territory, State, Local Government Organisations. No adverse comment or opposition to the conduct of the exercise was received during the briefings. These organisations provided comments which were incorporated into the final NOI to ensure that the requirements of the *Environment Protection (Impact of Proposals) Act 1974* are met through a process of consultation and negotiation.

The NOI includes the following:
a) the requirement for the proposed activity;
b) description of the proposed activity;
c) area of operations;
d) community consultation activities;
e) discussion of alternatives;
f) environment potentially affected by the proposal;
g) environmental protection safeguards.

The potential affects on the environment are detailed in terms of locations, habitats, infrastructure, land tenure and species. These are discussed in relation to the types of potential impacts that may occur including disturbance, damage, contamination, digging, fires and pest infestation. The protection safeguards are detailed in terms of responsibility at all levels to ensure environmental protection and the safeguards to be implemented, including: close liaison with land holders and local communities, prohibition of activities in defined 'no-go areas', protection of heritage sites, protection of sensitive habitats, waste disposal, prevention of soil erosion, livestock and native fauna protection and a number of other safeguards to address the issues raised during the community consultation process.

An environmental handbook will be issued that details the environmental protection safeguards to be implemented by all exercise participants. The process of community consultation will continue during and after the exercise. The shift in emphasis to the north will obviate an increase in these types of exercises and smaller scale activities. A close relationship based on mutual understanding needs to be maintained between Defence and the savanna community, not just for Kangaroo 95 but for all future exercises.

5. THE ADF'S ECONOMIC PRESENCE IN THE TROPICAL SAVANNAS

The economic impact of the ADF on the savannas is concentrated in the major centres, like Townsville and Darwin. The following summary of a recent economic analysis of the impact of the ADF on Townsville indi-

cates the contribution the ADF makes to the regional economy.

5.1 Case Example: The economic impact of the Australian Defence Force in the Townsville Region

A study by the Centre for Applied Economic Research and Analysis (Warren 1993) revealed *'significant flow-on effects to the local economy as a result of the presence of the ADF in Townsville'*. The study used a methodology based on Input-Output analysis. The flow-on effects were measured in terms of output of goods and services, value added, income generation and employment. The results are presented in summary form in Table 2.

Table 2: Impact of the ADF's presence in Townsville on the local economy in 1992-93

	Direct Impact	Flow-on Impact	Total Impact
Output($m)	226.82	282.88	509.7
Value Added ($m)	167.76	164.46	332.22
Incomes ($m)	163.85	84.57	248.42
Employment (no.)	4990	3600	8590

A similar story is expected to be found for the ADF presence in Darwin. The economic impact would be less in smaller regional centres, but it is expected that the impact would be a positive one on each community through an increase in economic stability and additional access to employment opportunities.

Defence recently initiated the Commercial Support Program (CSP). The main aim of CSP is to ensure that non-core support services and products are provided to core Defence activities in the most cost effective manner (Podger 1993). It will also enhance linkages between Defence and the private sector (Anon. 1991). This will improve the understanding of Defence needs and the cohesion required during both peacetime and periods of conflict. It also has significant economic impact on local commercial sectors as non-core service and product requirements of Defence are open to commercial tendering.

Other opportunities which will flow to the civilian community from enhanced ADF presence in the north are roadworks, airfield maintenance and catering (Ray 1993). Not only the permanent presence has an impact. There is also the impact of undertaking short term exercises involving ADF units from southern Australia. A recent example of this was the RAAF exercise 'Pitch Black' in 1991, when $215 000 was spent locally just on food (Ray 1993). Some of the more important construction activities recently commenced or planned for the next decade are: construction of RAAF Base Scherger, ongoing upgrade to RAAF Base Tindal, new administration complex for NORCOM and Lavarack Barracks Redevelopment Stages 1, 2, 3 and possibly further stages.

5.2 The ADF and mobility in the tropical savannas

The size of northern Australia and the sparse location of civilian and Defence infrastructure are key constraints to both an aggressor and the ADF. Overshadowing these are the dynamic seasonal extremes and innate variability of the savanna climate and its unforgiving terrain. These aspects drive the imperative to undertake periodic training throughout the savannas. Naturally this involves a parallel process of community wide consultation so that the primary objectives of the ADF are met without incurring an imposition on the rest of the community or a significant long term impact on the environment. In addition to training activity, there is a need to undertake research and development into enhancing the ADF's ability to traverse the tropical savannas as rapidly as possible.

Mobility is a key component in getting the ADF deployed. Research by CSIRO has played an important role in supporting work by Defence's Mobility and Terrain Analysis Group within the Engineering Development Establishment. CSIRO has been developing a cross country going rating expert system (XCGR) (Laut 1993). This is a trafficability prediction model for Army vehicles. The objective was to develop a method which will indicate the general range of going for a vehicle within an area using coarse readily available (i.e. not requiring precise field measurement) landscape data (Laut 1993). The expert system relied to a large extent on an irregular polygon based Geographic Information System (GIS) which incorporated spatial and site data on landscape attributes pertinent to trafficability.

A number of key issues were raised to facilitate future development of the XCGR. These issues are relevant to most other land users in the savanna:

a) The lack of landform (contour and relief), vegetation and soils data available for much of northern Australia at scales of 1:100 000 or larger;

b) Improved assumptions on lithology/slope/microrelief relationships;

c) Vehicle field performance based at first on field measurement of soil strength tests and vehicle trials;

d) Real time use of XCGR will require a better understanding of local, regional weather patterns and its impact on soils, stream flow and vegetation growth;

e) A better understanding of soil moisture dynamics;

f) Use of satellite imagery to provide a more up to date assessment of vegetation condition;

g) Increase in monitoring of the bio-physical environment to enhance the real time capabilities of the XCGR

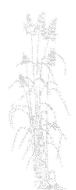

The issue of ground-based mobility has also been discussed by Dibb (1986). He concluded that to counter a protracted campaign of dispersed raids, we would need lightly armed ground forces, tactically mobile; with good communications, surveillance and reconnaissance capabilities. Carter (1993) details the three outcomes of these comments:

a) The permanent locating of the 2nd Cavalry Regiment at the newly developed Waler Barracks in Darwin;

b) The development and introduction of the 6x6 variant of the Landrover 110 Series which will provide enhanced mobility to the infantry Brigades and in the first instance the RSFUs;

c) The purchase of 112 wheeled armoured fighting vehicles (LAV25) the majority of which will be based in Darwin with 2nd Cavalry Regiment;

Ground-based mobility is also supported by the use of S70A Black Hawk battlefield helicopters and the future re-introduction of CH47D Chinooks at RAAF BASE Townsville (Carter 1993).

5.3 The ADF and the tropical savanna community

The ADF has a significant number of personnel either permanently or temporary present in the tropical savannas. Like all members of the community they are Australian citizens with the same or similar needs and expectations in terms of lifestyle and future for their children. This manifests itself in terms of the need for community services including housing, health, education and recreation. There is also the need for recognition that ADF personnel are involved in a role that is recognised by the Australian community as a positive and necessary part of Australian society. The ADF has undertaken reviews of the interaction of Defence and the civilian community and has recognised key issues in the tropical savannas that some community groups give particular importance to.

The relationship between the ADF and the civilian community has been comprehensively examined in the Wrigley (1990) report and the subsequent Interdepartmental Committee (IDC) report on the Wrigley review. Essentially there has been a recognition that Defence needs to undertake commercialisation/civilisation of a number of support functions, develop a Ready Reserve force, undertake a review into the relationship between civil and military authorities in the event of military conflict and that Defence undertake a public information program to better inform the public of Defence posture and planning considerations (Anon 1991).

At the local level there are community concerns over Defence's management of land used for training activity, particularly within the grazing industry and community conservation groups. As discussed previously public review of the draft EIS for TFTA highlighted a number of perceptions in the community relating to Defence's ability to appropriately manage land. The process of consultation in relation to TFTA and other training areas is not a once only activity. Defence intends to address community concerns within the framework of Environmental Management Plans.

Larger training areas will have consultative committees, that undertake informed discussion relating to issues raised by members of the public. Defence will be able to provide this committee with periodic updates on management issues such as pest management strategies, annual fire plans, etc. In addition, periodic independent environmental audits will allow further scrutiny of ADF management capabilities and identify opportunities for improvement.

5.4 The ADF and science and technology

As indicated briefly in the overview Defence has a significant internal capacity for scientific and technological support to the ADF. This is provided by the Defence Science and Technology Organisation (DSTO). The objective of DSTO is to *'enhance the security of Australia through the application of science and technology"* (Brabin-Smith 1993). Research is undertaken in four key areas; aeronautical, materials, electronics and surveillance. The work by Engineering Development Establishment detailed earlier is an example of this which is relevant to the savanna region.

When the ADF has research needs not specifically related to military objectives, the ADF relies on CSIRO and other organisations. For example, support for the effective land management of Defence controlled property is obtained through CSIRO at a cost $0.5 m annually. This involves specialist environmental research and analysis projects done by CSIRO staff with direct benefit to Defence. The two current projects already discussed are examples of Defence seeking to apply modern technology to meet environmental management objectives. The implementation of DTAMIS is expected to place the ADF in a position to further enhance management of the land it controls. In addition, the development and implementation of modern methods of land condition analysis and monitoring will ensure that the ADF is an effective land management organisation.

No landholder has all the answers to the management issues facing them in the tropical savannas. However, as outlined above Defence has emplaced a framework of environmental management to address its obligation to ecologically sustainable development. In addition, Defence is responsive to community expectations and the increasing knowledge provided from both Commonwealth and State/Territory research organisations.

An example of this is current research by the CSIRO Division of Tropical Crops and Pastures on managing tropical woodlands for grazing and conservation and coastal zone management and research by the Division of Soils on minesite rehabilitation, rural land use

impact on catchment-estuary water quality and soil degradation in the semi-arid tropics: assessment, processes and risk prediction.

5.5 The ADF and the future

This paper contains some of the important aspects of the ADF's intentions of enhancing its presence in the tropical savannas. The most obvious task is to continue the shift in focus to the north for the Army and the Air Force and west for the Navy. More specifically this will involve:

- Completion of RAAF Base Scherger, near Weipa;
- Completion of stage three of the development of RAAF Base Tindal;
- Completion of 2nd Cavalry Regiment equipment upgrade;
- Relocation of a Brigade to the north by the end of the decade, including an armoured regiment with one Tank Squadron;
- Increase the number of the Army's major exercises undertaken in the north;
- Further development towards basing half the Navy's major surface combatants at HMAS Stirling in Western Australia;
- Development of Townsville Field Training Area;
- Commence development of the North Australian Training Area;
- 161 Reconnaissance Squadron to relocate to RAAF Base Darwin in 1994-95;
- The remainder of 1st Brigade to complete the move to Darwin by the end of the decade.

The projects listed involve or will involve major capital works (Ray 1993). This will be matched by an increase in flow-on effects due to increased Defence employment and project funds in the north. An important part of this will be enhanced integration with the civilian community through such activities as CSP and Ready Reserve.

The use of the tropical savannas for training and exercise purposes will be appropriately managed on the basis of enhancing the existing environmental management framework to meet the objectives of the Commonwealths national strategy for ecologically sustainable development. At the local level this will occur through implementation of environmental management plans for Defence training areas and training activities. Applied land management research will continue to be supported by Defence as will the incorporation of newly developed land management techniques and technological advances.

The process of increasing Defence personnel in the tropical savannas will necessitate a strengthening of linkages between Defence and the civilian community. There must be mutual understanding of the need for, and commitment to, integration of Defence personnel within the wider community as this will be the backbone of Australia's policy of 'self-reliance'.

References

ANAO (1992). *Department of Defence, Management of Military Training Areas, Efficiency Audit*. The Auditor General, Audit Report No. 38, 1991-92. AGPS, Canberra.

Anon. (1976). *Australian Defence*. Presented to Parliament by the Minister for Defence, the Hon D. J. Killen, November 1976. AGPS, Canberra.

Anon. (1987). *The Defence of Australia*. Presented to Parliament by the Minister for Defence, the Hon K. C. Beazley. AGPS, Canberra.

Anon. (1991). *The Defence Force and the Community*. Report of the Interdepartmental Committee (IDC) on the Wrigley Review. AGPS, Canberra.

Anon. (1993). *Strategic Review 1993*. AGPS, Canberra.

Ayers, A. J. and Beaumont, A.L. (eds) (1993). *Defence Report 1992-93*. AGPS, Canberra.

Ayers, A. J. and Gration, P. C. (1991a). *Force Structure Review*. AGPS, Canberra.

Ayers, A.J. and Gration, P. C. (1991b). *Evironment and Heritage Protection*. Department of Defence, Canberra.

Brabin-Smith, R. G. (1993). Program Eight: Science and Technology. In *Defence Report 1992-93*, (eds. Ayers, A. J. and Beaumont, A.L.), pp. 135-51. AGPS, Canberra.

Carter, G. (1993). New developments in land forces: The impact on operations in the defence of Australia. In: *The Army and the Future: Land Forces in Australia and South-East Asia*, (ed Horner, D.), pp. 259-76. AGPS, Canberra.

Cooksey, R. J. (1988). *Review of Australia's Defence Facilities*. Report to the Minister for Defence. AGPS, Canberra.

Dames and Moore (1993). *Draft Environmental Impact Statement: Proposed Townsville Field Training Area*. Department of Defence, Canberra.

Dibb, P. (1986). *Review of Australia's Defence Capabilities*. Report to the Minister for Defence. AGPS, Canberra.

Grey, J.C. (1993). Program Three: Army. In: *Defence Report 1992-93*, (eds Ayers, A. J. and Beaumont, A.L.) pp. 47-61. AGPS, Canberra.

Laut, P. (1993). *Cross Country Going Rating Expert System: A Trafficability Model for Army Vehicles'* CSIRO, Division of Water Resources, Divisional Report 93/5. CSIRO, Canberra.

Lewis Young, P. (1994). Defending Australia's North. *Asian Defence Journal*, No. 3, March: 14-23.

Podger, A. S. (1993). Program Six: Acquisition and Logistics. In: *Defence Report 1992-93*, (eds Ayers, A. J. and Beaumont, A.L.), pp. 99-121. AGPS, Canberra.

Ray, R. (1993). The strategic imperative of the move to the North. In: *The Army and the Future: Land Forces in Australia and South-East Asia*, (ed Horner, D.), pp.229-38. AGPS, Canberra.

Warren, J. (1993). *The Economic Impact of the Australian Defence Forces in the Townsville Region*. Centre for Applied Economic Research and Analysis, JCU, Townsville.

Wrigley, A. (1990). *The Defence Force and the Community, A Partnership in Australia's Defence*. AGPS, Canberra.

Chapter 8

Sustainable Mining in Australia's Tropical Savannas

Geoffrey Ewing

*Assistant Director, Australian Mining Industry Council, Dickson, ACT 2602
Australia*

Abstract

A significant part of the Australian mining industry's activity occurs in tropical savannas. Major operations include the central Queensland coal mines, the bauxite mines at Weipa and Gove with their associated refineries and smelters, the Mt. Isa region, the gold and base metal operations at Charters Towers, the undeveloped projects in the Carpentaria block, and the Ranger uranium, Groote Eylandt manganese and Argyle diamond mines.

The industry wants to continue to contribute to the sustainable development of Australia's tropical savannas by exploring for, and mining deposits in an environmentally acceptable manner. It has made a significant contribution in the past and wishes to do so in the future provided the land access policy framework in Australia allows this to occur.

Despite the wealth, employment, environmental knowledge and infrastructure generated by the industry in this region, the industry finds itself being prevented from exploring in significant parts of Australia's tropical savannas due to an increase in the amount of lands allocated to conservation and aboriginal use.

A key requirement of future policy frameworks to enable the continued growth of Australia's mining industry must be the recognition by industry, governments and the community that the environment and development must be integrated and that lands need to be properly managed.

1. INTRODUCTION

A significant part of the Australian mining industry's activity occurs in tropical savannas. Most mines in the country operate pursuant to mining leases granted over pastoral leases and many of those leases are over savannas. Major operations include the central Queensland coal mines, the bauxite mines at Weipa and Gove, the Mt. Isa mines, the gold and base metal operations at Charters Towers, the undeveloped projects in the Carpentaria block, and the Ranger uranium, Groote Eylandt manganese and Argyle diamond mines.

The industry wants to continue to contribute to the sustainable development in Australia's tropical savannas by exploring for, and mining deposits in an environmentally acceptable manner. It has made a significant contribution to both economic growth and environmental science in this country in the past and wishes to do so in the future. This will only continue if the land access policy framework in Australia allows it.

While the industry is justifiably proud of its record in recent years, the theory and practice of continuous improvement can assist not only the scientific approach to environmental care but also by ensuring that the community as a whole becomes increasingly aware of the role that the mining industry is playing in our society. This community education focus must run the full gamut of the industry involvement in our social structure, from the economic benefits of mining, to the contribution that our industry is making to science in this country, both at the industry technological level as well as in the environmental arena. Environmental considerations have been integrated into industry planning not only because it makes good economic sense, but because the industry is well aware that unless the community is satisfied as to its environmental performance, it will not be able to maintain existing resource development, much less expand in a way which will allow the continuance of our high standard of social infrastructure.

2. SUSTAINABLE DEVELOPMENT

'Sustainable development' was defined by the World Commission on Environment and Development (1987) as ensuring *"that it meets the needs of the present without compromising the ability of future generations to meet their own needs."*

The greater the efficiency of the economy, the greater is the ability of the economy to improve the material well-being of the community, generate the capital to improve non-material well-being and at the same time ensure that the environment is protected.

The mining industry is the best key we have to unlock the barriers to our continued material and non-material prosperity and to do so in a responsible and sustainable manner.

As the editorial in *The Courier Mail* headed 'Still a prime nation builder' stated on 7 July 1994:

'... however successful we are in developing sustainable export-oriented manufactures and services, our natural resources provide huge wealth and the promise of a secure future.'

A 10% increase in the value of our export earnings from mining alone would realise an additional $3 billion to the economy. It will take a lot of new enterprises and industries in other areas to realise that sort of benefit.

Sustainable development requires that growth in world demand be met as efficiently as possible, while at the same time minimising the environmental impact. Australia is in a unique position and has a responsibility to assist in meeting this demand. Not only is the Australian minerals industry efficient, its environmental impact is relatively low compared with other production sources.

Mining operations directly impact on some 0.02% of our land area. A large number of these operations occur on savanna areas although they disturb but a tiny fraction of the total savanna.

Provided that a balanced, sensible approach is adopted, the goals of sustainable development are achievable. It is possible for Australians to continue to enjoy rising material living standards and for exports to satisfy the needs of foreign customers and repay our burgeoning foreign debt, while at the same time enhancing the quality of the environment. The key is increasing the productivity with which our assets, labour, physical capital and natural capital are utilised. However, inappropriate policies will result in both sharply reduced living standards and deterioration of environmental conditions.

The Ecologically Sustainable Development Working Group on Mining (1991) said in its final report:

'...the mining industry has had less broad impact on the Australian environment than some economically comparable sectors of the Australian economy, even though in some cases it has had extensive and intensive impacts. Modern environmental impact assessment can anticipate, and management practices including mine site rehabilitation and can avoid or minimise, adverse impacts in most circumstances.'

A member of the ESD Group on mining, Dr Ian Gould of CRA, said in a paper at the Australian Financial Review Conference in Sydney on 9 April 1992:

'The principles of sustainable development are being adopted worldwide. They embody sensible approaches to conservation and development. What is different about Australia is the belief by some in the community that there are those who are happy to enjoy the standard of living for which the mining industry is substantially responsible, but

are unwilling to recognise the need to continue developing our mineral resources. There are no free lunches — we cannot debate orebodies into more convenient locations or legislate national prosperity. Enjoying the use of mineral products means facing up to the issues involved in providing them.

2.1 The Economic & Social Legacy of Mining

The mining industry is an industry of vital importance for all Australians. According to the Australian Bureau of Agricultural Resource Economics (1993), the mining industry contributed some AUD$30 billion to our export earnings in 1992/93, representing some 50% of merchandise exports. Some AUD$12 billion of Australia's manufactured exports including manufactured metal products, cars, paints, plastics, are dependent on mineral products. Mining directly employs 2.1% of the total workforce and 9.4% of total employment is dependent on Australia's metals and minerals production. The industry paid almost AUD$3.5 billion in taxes to governments in 1992/93 and in the same period the minerals resources sector contributed some 23% of new private enterprise capital expenditure.

It is quite clear that it is the developed countries which have taken and continue to take most care of the environment. The third world nations have neither the cultural impetus nor the short-term financial and technical capability in many cases to ensure the same level of standards of environmental rigour as do the developed nations. Further, it has not been recognised by the less developed countries in many instances that it is economically more advantageous to inbuild environmental protection measures as the most cost-effective process in the longer term. This problem is one which is being addressed by these countries but change will be evolutionary rather than dramatic.

Stronger economies mean that we have the ability to develop longer term and more efficient environmental and other strategies that may produce higher costs in the short term but greater efficiencies and economies for more durable and effective outcomes in the future.

Success in a competitive world breeds innovation through the desire to stay in front. We have developed a commitment to excellence and continuous improvement in all facets of our operations from the technical to the management side where we are acknowledged as market leaders. The Australian mining industry enjoys a world-wide reputation as a leader in environmental management and rehabilitation. It carries out a significant proportion of environmental research in this country and is Australia's largest employer of environmental scientists in specialist fields such as forestry, botany, zoology, hydrology, air quality, meteorology and microbiology. Although most communities pay lip service, at least to the industry contribution in this

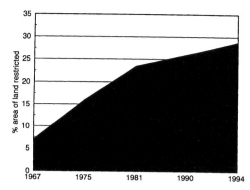

Figure 1. Area of Australia with restricted land access for exploration and mining.

area, our reputation in the public domain is a far less happy one than it might be.

As *The Courier Mail* (7 July 1994, editorial) said in the previously quoted editorial, in referring to some Brisbane secondary school teachers and their attitude to the mining industry: '*is, presumably, seen by them as a despoiler of the environment and a threat to the warm, fuzzy and fundamentally illogical view of the world that has become so popular in some sections of the community.*'

The challenge for our community, including governments, the press, academia and industry is to ensure that there is widespread appreciation of the economic and broader social importance of the mining industry.

2.2 Exploration

Only about 1 in 1000 exploration prospects ever becomes a mine. Undertaking exploration is therefore a very risky exercise.

Most of the exploration being undertaken in Australia at the moment is carried out by about ten major mining houses. The ten leaders in the field are all truly international companies and are exploring and running operating mines in other parts of the world. These large companies usually operate their exploration on something akin to a 5-year plan. In other words, exploration which is being undertaken at the moment was most likely decided upon by these companies in 1989 or 1990.

In making the decision to explore, these companies look at three major factors:
- geological potential;
- the fiscal and taxation regime which applies in the country concerned; and
- administrative ease of obtaining approvals and sovereign risk.

Australia has by and large been very competitive in international terms in all of these three elements. The geological and fiscal factors remain competitive in world terms but the administrative/sovereign risk regimes which now apply as a result of land access

considerations such as the *Native Title Act*, World Heritage issues and restrictions based on perceived environmental issues are leading to a marked shifting of the balance which has favoured Australia in the past.

Additional impediments in the form of delay and significant expense in gaining approvals provide an incentive for explorers and miners to put their exploration dollars elsewhere.

Australia is a world leader in exploration and mining technology and Australian companies do and will find it increasingly easy to move larger portions of their exploration budgets offshore if life is made more difficult at home.

Once these exploration dollars move offshore, it is very difficult to bring them back to ensure continued mining development in this country.

The application of strict single land use regimes over national parks, conservation reserves and other environmentally sensitive areas constitutes a major economic problem for the community. A single use land regime over national parks, conservation reserves and wilderness areas precludes even the assessment of mineral potential in an area. Consequently governments face decisions about these areas, often as a result of single-issue group pressure, without adequate information on the resource base and without appropriate assessment of the development and conservation alternatives. The Commonwealth Government *One Nation* statement of February 1992 committed to resource assessment of all areas but action has not followed the words.

The area of land being restricted or closed off to new mineral exploration or mining activities is increasing, but at a slower rate than in the past (Figure 1). In 1967, approximately 7% of Australia was restricted or closed to exploration and mining. By 1975, this figure was 16%. In 1981 this figure had grown to 23.5% and by 1990 to 26% with the most recent figures indicating it is 28.8% of the country. The current frighteningly large figure of 28.8% represents an increase of 311% over the 1967 figure of 7% in a period of 27 years. Most of this restricted land comprises nature conservation reserves and land held under Aboriginal control. The figure of 28.8% does *not* take into account native title claims.

Restricting (not just banning) exploration and mining in a large number of reserves around the country and the prescriptive, expensive and cumbersome provisions of the *Native Title Act 1993* have all contributed to making Australia a less attractive place to explore. Unwarranted additional restrictions on exploration in valuable yet plentiful areas such as our tropical savannas will only add to a growing range of impediments which are making Australia a less attractive place to explore than it has been in the past.

Professor Alban Lynch in a paper in the June 1993 edition of Mining Review said:

'Industrial development during the next forty years will be dominated by three issues, world population, computer-based technology and the environment. The demands on the mining industry will be for more minerals to provide materials and energy for the growing population, lower product prices because of increased competition, and a high standard of environmental management.'

Success will depend on making best use of the technologies which are currently available and developing new technologies when necessary. This requires many skills of the highest quality. In fact the future of the Australian mineral industry might well be defined by the relationships:

PROSPERITY = ORE + SKILLS - OTHER COSTS
SKILLS = EDUCATION + EXPERIENCE

There remains a common myth that one can carry out a resource assessment of a particular area today, and tomorrow make a decision for all time that the social balance lies in favour of preventing either exploration or mining in the area. This approach is a fallacy as developing techniques and fluctuating metal prices mean that environmental impact weighting and financial considerations can change dramatically in a short time.

We can explore these days with minimal impact and there is still a healthy measure of luck in all new 'finds'. The Century zinc/lead deposit north of Mt. Isa was discovered only a few years ago in what was one of the most explored areas in the country over the last 100 years. It is a deposit which contains some 10% of the world's known zinc resources.

We are becoming far better at exploration — airborne geophysical and geochemical exploration means that exploration can often be totally non-impactive or have only minimum physical impact. Future exploration will increasingly be directed to the identification of deeper deposits which will result in less surface disturbance.

Exploration, whether successful or unsuccessful, equates to knowledge gained. The 'book-burning' mentality of the type displayed by the recent Commission of Inquiry into the most appropriate use of Defence land at Shoalwater Bay is mind-boggling. The Commission there essentially indicated that there should be no exploration in case something was found!

The knowledge to be gained from exploration is vital so that proper land care decisions may be made.

3. COMMUNITY PARTICIPATION AND AVOIDING CONFLICT

It is all very well to point out the fallacies in the arguments of those who have a different view to one's own but if we are to optimise the social, environmental and economic returns from all of our land, including our

tropical savanna, we must work on resolving conflict in the most efficacious way.

Direct approaches to any expected opponent may be the best approach. Too often companies feel oppressed by the weight of existing regulation and procedures that they tend not to look outside that framework — this can be exacerbated by an often well deserved fear that speaking to a potential adversary may lead to legitimising another process or player and hence complicating the processes even further — the alternative of 'letting sleeping dogs lie' is often preferred. One of the difficulties in seeking out the likely opposition is in determining which group to deal with. Local environmental groups, state groups, or national groups — often disagree amongst themselves on the issue and on the problem and identifying the real issue in dispute can be very difficult.

Relationships with local communities are an integral part of modern mining projects. Appropriate strategies are established with the same priority as those of our other core components of any project, such as technical, economic and human resources.

At an AMIC Community Relations Workshop in Adelaide in May 1990, Mr George Gauci of Canning Resources Pty Limited identified some of the matters which need to be addressed in establishing our strategies for better and more effective communication:

'*The target. Who are the stakeholders within the community that should be included in the exercise? Are there social areas as well as geographic areas to be considered?*
The timing. At what stage of the project development do we approach local communities to discuss our proposals? If contact is made too early sufficient information to satisfy the concerns of the community may not be available, and consequently arouse suspicions as to our intentions. On the other hand, if one delays the contact until the data is available it may be too late to prevent perceived concerns becoming real issues.
The process. What level of involvement of the community are we seeking? It could be:

— *a one-way flow of information;*

— *two-way communication involving public meetings, liaison committees and public displays;*

— *full involvement of local communities in the solution to issues. The risks involved are that it can raise expectations that cannot commercially be met, and there may be some commercial benefit to some people, such as advance information on land, housing or services.*
Impacts of increased population. Most projects result in the introduction of people into a region that will have social implications for the established and newly introduced population.
The social success of such a move depends on reactions of the existing community, the nature of the incoming population and their ease of adoption, and the degree of interaction between the two.

Role of senior management. How often should senior executives visit the area to show their support? The seniority of staff permanently resident in the community in early development and ongoing operations needs to be addressed.'

All of these issues are ones which must be addressed by all participants in a debate on developing plans for appropriate land use.

The implications of World Heritage listing or nomination on development.

One of the other impediments to exploration and mining in Australia in recent years has been the addition of natural areas to the World Heritage List. Exploration and mining are not prevented by listing but the practice has been in Australia when properties are listed that additional controls are instituted which effectively prohibit mineral resource development and even exploration.

The criteria for World Heritage Listing of natural areas are that they must:
(i) be an outstanding example representing the major stages of the earth's evolutionary history;
(ii) be an outstanding example representing ongoing geological processes, biological evolution and man's interaction with his natural environment;
(iii) contain superlative natural phenomena, formations or features; and
(iv) contain the most important and significant natural habitats where threatened species of animals or plants of outstanding universal value live;

Probably most of the world's 411 World Heritage listings are cultural and/or architectural properties. Australia's ten listings are all natural areas: Wet Tropics of Queensland; Fraser Island; Kakadu National Park; Shark Bay; Willandra Lakes Region; Lord Howe Island Group; Great Barrier Reef; Australian East Coast Temperate and Sub-Tropical Rainforest Parks; Tasmania Wilderness; Uluru National Park. Kakadu lies entirely within the tropical savannas while the Queensland Wet Tropics listings contains significant areas of tropical savanna.

It is not suggested that we should not have some of our marvellous natural areas on the World Heritage List. However, there is no logic in a total ban on exploration and mining in those areas.

The plummeting of rural property values at Willandra Lakes in NSW over the past ten years, due to World Heritage Listing and inept management by the Commonwealth Government, is a stark lesson for us all.

As former Commonwealth Environment Minister Barry Cohen (1993) said of the Willandra Lakes listing in 1981:
'*The impact on potential purchasers was devastating. Farming was tough enough without acquiring a property that looked like the maze at Kew Gardens. Buyers stayed*

away in droves and when a property finally sold it did so at less than half the value of similar properties outside the region. Small wonder farmers were angry when told by government the drop in their property values was due to the rural recession. Hardly surprising also that Lake Eyre property holders were terrified they would suffer the same fate.'

The proposed nomination of Lake Eyre will be controversial for some time. The Commonwealth Government has not delineated the area it is contemplating for nomination but Environment Minister Faulkner (1994) recently indicated *'we are pursuing an assessment of the values of Lake Eyre.'*

The Wet Tropics of Queensland World Heritage Area was listed in 1988. The Wet Tropics area covers about 9003 sq. kms in a narrow 420 km band from Townsville to Cooktown. To quote from a socio-economic study on the Wet Tropics area prepared the Queensland Government in 1988:
'Over 7.5% of Queensland's population lives within 50 km of the World Heritage Boundary. This area of North Queensland is one of the continent's most rapidly expanding population areas. Immediately adjacent to the proposed boundaries are some of Australia's largest tourism developments. Indeed it is the success of the region in the tourism field that poses some of the greatest challenges and ultimate conflicts with the proposed World Heritage Boundaries' (Queensland 1990).

The Queensland Government's report pointed out that the north-east Queensland area includes some of the most prospective areas for gold, tin and tungsten resource development in the State, and possibly in Australia. Based upon the Kakadu Stage II World Heritage decisions of the Commonwealth Government, the mining industry does not view future mineral exploration and development within the Wet Tropics Heritage Area with any degree of confidence. The Commonwealth Government has stated mining could continue in the nominated area provided World Heritage values are not impacted. (It was confirmed by the Federal Minister for the Environment at the time, Senator Richardson, that intending miners might not be allowed to remove trees which would effectively prevent mining). In fact, of the 30 mining leases and claims in existence at the time of nomination in 1988, only about half of these now exist. It is unlikely that any exploration or mining leases will be granted in the area in the future. This is not because every bit of the World Heritage Area would be changed by exploration or mining: it is simply because it is in the nominated zone.

Cairns and Townsville are major regional population centres in Australia. They require the normal infrastructure support for major centres such as water, sewerage, electricity and roadways. Because of the proximity of the boundaries of the World Heritage Area these normal infrastructure requirements could still undergo major cost escalations due to Park boundaries and regulations.

Six years on from listing, the management plan for the area is yet to be finalised with consequent uncertainty for everyone.

Guidelines issued by the Commonwealth of Australia on 11 December 1987 relating to the proposed World Heritage Area in North Queensland stated:
'Grazing of stock can continue, as can mining operations, and the provision of infrastructure like roads and electricity and water supply systems, PROVIDED SUCH ACTIVITIES DO NOT INVOLVE CLEARANCE OF AREAS OF RAINFORESTS OR OTHER THREATS TO WORLD HERITAGE VALUES'.

The pastoral leases in the area which were unequivocally to be unaffected by the nomination are now in question.

Current electric power line upgrades into Cairns and Port Douglas would appear to be prohibited under the guidelines as would water supply upgrades, roadway improvements, and other necessary infrastructure projects. Although no decision has been made, it appears unlikely the Tully Power Station will proceed. The Australian World Heritage Guidelines are not intended for urbanised areas yet the boundaries surround some of the country's fastest growing areas. The conflict between these two facts cannot be understated or ignored.

We must be realistic and pragmatic in our approach to sustainable development. World Heritage listing in Australia has meant in every case a negative effect on resource development. The decisions have usually been made without a proper assessment of the economic effect on the community, on industry or on individuals.

If we see further World Heritage Areas created, including our savanna areas, without a realistic assessment of the consequences, the economic cost could be very great indeed.

It must be emphasised that it is not a mining industry view that economic considerations must always prevail over proper social and environmental considerations. The issue is all about balance and recent trends in this country indicate that we very often get the balance wrong.

4. COMMUNICATING THE GOOD RECORD

The mining industry has undertaken many exercises in communicating its environmental expertise and much of the work done relates or applies to the savannas. The initiatives include:

i) **Australian Minerals and Energy Environment Foundation (AMEEF)**

In 1991, this Foundation was established to promote the implementation of sustainable development principles in Australia's mineral and energy related industry.

This institution aims to develop a repository of environmental knowledge and to offer practical advice on our own environmental management.

It plans to develop databases on environmental aspects of minerals and energy and to promote community education and the transfer of environmental expertise.

ii AMIC Annual Environmental Workshop

In October 1994, the Australian Mining Industry Council will hold its nineteenth Annual Environmental Workshop. These workshops have contributed handsomely to the sharing of knowledge throughout the industry on best practice in mine site environmental management and rehabilitation. These workshops have for many years been attended by government Mines and Environment Departments and semi-government organisations with an interest and responsibility in environmental science such as CSIRO.

iii AMIC Minesite Rehabilitation Handbook

AMIC recently published (Ward 1990) a *Minesite Rehabilitation Handbook* which has the approval of the Department of Mines in every State. This is used extensively by environmental practitioners both within and outside the industry. This document has now been translated into Indonesian and is also being translated for appropriate use in China and on the African continent.

iv Australian Centre for Minesite Rehabilitation

This Centre has been established in Queensland with the support of AMIC. It is a joint venture between the Australian Minerals Industry Research Association and three of the major groups in Australia performing minesite rehabilitation research, CSIRO, University of Queensland Centre for Minesite Rehabilitation and the Curtin University of Technology Mulga Research Centre. Its vision is for the development of sustainable land systems for rehabilitated minesites acceptable to industry, government and to the community.

These initiatives are all very constructive and useful but their existence and usefulness must be brought to the attention of a much greater number of Australians if they are to be fully effective.

4.1 Rangelands

The Commonwealth Government has initiated a national rangelands study to develop a *National Strategy for Rangeland Management and Action Plan*.

The area of rangelands as defined by the Australian Bureau of Agricultural Resource Economics covers most of the continent and obviously includes most of our savannas.

In an Issues Paper on Rangelands released in February 1994, produced by the Australian and New Zealand Environment and Conservation Council and the Agriculture and Resource Management Council of Australia and New Zealand, it is suggested:

'The *mining industry has had considerable success in adapting to terms of trade pressures, through productivity improvements related to management, technology, marketing and by choosing the most appropriate ore bodies to mine within an economic framework. Environmental costs, as prescribed by legislation, are now considered in the economic assessment of an operation from the outset, unlike in pastoralism where such costs may only be considered after degradation has occurred. For mining, it may be worth considering whether the industry is meeting the full environmental costs of resource exploitation.*'

While there is no indication that anything but good will come out of the Rangelands process, it is yet another inquiry which all industry groups need to be involved in to ensure proper understanding of their interests. Many groups have expressed dismay at the plethora of inquiries which make doing business in this country that much harder.

5. Conclusion

Australia has a unique natural heritage which we should all be doing our utmost to conserve.

Mining and the environment are not intrinsically in conflict and our mining industry in Australia has contributed far more to scientific excellence in terms of understanding of our physical environment than any comparable industry in the world.

Despite the wealth, employment, environmental knowledge and infrastructure generated by the industry in this country the industry finds itself being prevented or restricted from exploring in significant parts of Australia, including its tropical savannas, due to an increase in the amount of lands allocated to conservation and Aboriginal use.

A key requirement of future policy frameworks to enable the continued growth of Australia's mining industry must be the recognition by industry, governments and the community that the environment and development can and must be integrated in a sustainable way.

We are well on the way to doing this but continuous improvement in all aspects of our land access approach is vital if our standard of living is to be maintained.

References

ABARE (1993). Commodity overview. *Agriculture and Resources Quarterly*, 5: 457-463.

Cohen, B. (1993). Lake farmers adrift on a sea of uncertainty. *Australian*, 3 August, pp. 11.

Courier Mail (1994). Mining: Still a prime nation builder. Editorial, 7 July, pp. 8.

Ecologically Sustainable Development Mining Working Group (1991). *Final Report*. AGPS, Canberra.

Faulkner, J. (1994). Speech at the launch of the "State of the Environment Reporting: Framework for Australia., 28 June.

Gauci, G. (1990). Corporate strategies and community relations. *Mining Review*, November, pp. 34-35.

Gould, I. (1992). Industry response to ESD: mining and minerals processing. AFR Conference: The Cost of ESD to Australian Industry, Sydney, 9 April, pp. 6.

Lynch, A. (1993). Maximising skills to increase prosperity in Australia's mineral industries. *Mining Review*, **17**: 3-5.

Queensland. Premier's Department. (1990). *Socio-economic study wet tropical rainforests North Queensland*. Cameron McNamara, Brisbane.

Ward, S. (1990). *Mine Rehabilitation Handbook*. Australian Mining Industry Council, Dickson, ACT.

World Commission on Environment and Development (1987). *Our Common Future*. Oxford University Press, New York.

Chapter 9

Aboriginal people of the Tropical Savannas: Resource utilization and conflict resolution

Darryl Pearce[1],
Andrew Jackson[1],
and Richard Braithwaite[2]

[1] Northern Land Council,
Darwin, NT
Australia

[2] CSIRO Division of Wildlife & Ecology,
Tropical Ecosystems Research Centre,
Private Bag No 44,
Winnellie, NT 0821
Australia

Abstract

From the beginning of Australia's colonisation by Europeans, Aboriginal people have struggled physically, politically and through legal processes to gain recognition of their right to own and manage their traditional lands, seas and resources. Though often forcibly moved and resettled long distances from their homeland, most Aboriginal people have never relinquished their attachment to their traditional country nor lost interest in being reunited with it.

This paper aims to contribute to the understanding of non-Aboriginal people about why land or 'country' continues to be of such significance to contemporary Aboriginal society, how and why Aboriginal people contribute to the development and utilization of the tropical savannas and how apparent and real conflict between Aboriginal and non-Aboriginal users of the savannas can be resolved.

We examine the historical context which has determined where Aboriginal people now find themselves and considers how and why Aboriginal people have traditionally used the savannas for tens of thousands of years. We also look at the current economic situation of Aboriginal people in the savannas and why they are such important contributors to the region. Finally, we look at areas where there are conflicts, and propose some relatively simple ways to reduce conflict between Aboriginal and non-Aboriginal users of the savannas. Communication, mutual understanding and respect are the key to a more harmonious and prosperous future.

1. INTRODUCTION

This is an attempt to present, in a succinct way, an Aboriginal view of the world of the savannas. We recognise that much of the information presented has not been available to the general public, nor even many of the scientific community in the past.

In our view, this is a significant part of the problem, that is, the real and apparent conflict that arises between the original owners/managers of the savannas, Aboriginal people and the more recently arrived non-Aboriginal settlers. It is our intention to provide a reasonably detailed understanding for non-Aboriginal people about why land or 'country' continues to be of such significance to contemporary Aboriginal society, how and why Aboriginal people contribute to the development and utilization of the tropical savannas, and how apparent and real conflict between Aboriginal and non-Aboriginal users of the savannas can be resolved.

1.1 Understanding Aboriginal views about land

Smyth (1994) in *Understanding Country*, describes how Aboriginal people view their land, their 'country'. For Aboriginal people the significance of land and sea is inextricably linked to their creation beliefs which surround the origins of the landscapes and seascapes, and the animals, plants and people that inhabit them.

Aboriginal peoples' creation beliefs vary greatly from region to region, but they generally describe the journeys of ancestral beings, often giant animals or people, over what began as a featureless domain. Mountains, rivers, waterholes, animal and plant species, and other natural and cultural resources came into being as a result of events which took place during these dreamtime journeys. These stories explain the presence of all features of the Australian landscape.

An important feature of these beliefs is the physical manifestation of important points in these travels which became known in non-Aboriginal terms as 'sacred sites'. These sites can vary enormously in size from such things as a tree to an area thousands of square kilometres. They are the physical link between the stories and the land.

Sacred sites are fundamentally different to such things as the construction of churches, which are a display of Christians' respect for God. Churches are not, in themselves, the physical manifestation of God him or herself, but rather a place for his or her worship, and a token of Christians' esteem for God.

While Aboriginal beliefs are probably no more or less provable than creationism, Darwinism or any other way of explaining why the world is the way it is, the fundamental fact is that these views must be respected if we are to succeed as a coherent society. They are a way of making sense of the world, of explaining how the world came to be as it is. Most importantly, whether they are a system of myths, beliefs or so-called scientific fact, they are a way of providing a solid base for a societal group to develop, mature and order their lives. In fact, they may be a more effective way of giving social unity than European law and order or science.

Aboriginal and Torres Strait Islander peoples belong to diverse, contemporary communities, each containing individuals with different perspectives, life experiences and aspirations. While there are many shared interests based on their status as indigenous Australians, it must be understood that there is a diversity of opinion within communities about all issues, including the significance of land and sea.

1.2 What Is 'Country'?

In this context, 'country' means place of origin, literally, culturally or spiritually. It can have the political meaning of 'nation', but refers to a clan or tribal area rather than a nation-state such as Australia. 'Country' refers to more than just a geographical area, it is a shorthand for all the values, places, resources, stories, and cultural obligations associated with that geographical area. For coastal Aboriginal peoples and Torres Strait Islanders, 'country' includes both land and sea areas, which are regarded as inseparable from each other.

1.3 Owning And Caring For Country

Systems of ownership of, access to and responsibility for traditional Aboriginal clan estates differ from place to place, but there are some common elements which indicate the importance of particular areas to particular people. Membership of a clan, and hence an association with that clan's country, is given at birth.

Sons and daughters retain clan membership for life, even though they may move away and live on other clan estates (such as a husband's or wife's clan country), or into community settlements or towns.

Clan membership provides access rights to the hunting, fishing and gathering resources of the clan estates, and often also some rights to resources on other related estates.

Its members are also responsible for carrying out ceremonies, observing various taboos (such as restrictions on who can eat and prepare certain foods) and for physically managing the estate's resources (such as by burning the country in the appropriate manner).

It was this inherited association with a particular country and its sacred sites, dreaming tracks, stories, totems and other features, which in precolonial Australia provided Aboriginal peoples with their individual and group identities. The severing of that relationship to a particular country as happened across much of Australia during the colonial period and into recent times, denied Aboriginal people a place in their

kinship system, access to resources and their basis of spiritual belief.

The importance of maintaining a connection with their traditional country continues to be of fundamental importance for Aboriginal and Torres Strait Islander peoples' identity and well-being across much of Australia today.

Contrast this with the view of non-Aboriginal Australians, who while they may develop an historical attachment to land by virtue of birthright or inheritance, always retain a sense of ownership or title rather than a sense of belonging in the comprehensive Aboriginal sense.

Non-Aboriginal people generally form a bond with the land that has an economic focus. Families can and often do own properties for generations and then sell them if the economic imperative is there.

To Aboriginal people, it is extraordinary that European migrants could move to Australia, buy property and then subsequently sell it for a profit, sometimes after a couple of generations and move back to Europe, but not necessarily to the place of their birth.

This intrinsic difference between land as a tradeable commodity and land as a central focus of your culture is the principal, although not only, differentiation between Aboriginal and non-Aboriginal people.

2. THE DIVERSITY OF ABORIGINAL PEOPLE

Like any other group in Australian society, Aboriginal people do not form a homogeneous unit, rather, they are distinct individuals with greatly varying needs, aspirations, backgrounds and skills.

The regional, social and cultural variations within Aboriginal groups are, in fact, considerably more diverse and complex than is the case for many other cultures.

A good example of this is the difference between Aboriginal people of the arid zone and people of the tropics. Arid zone people tend to be slighter of build, darker of colour and have significantly larger areas of country in which they hunted and foraged. This, in turn, is something which is certain to affect their outlook in various ways.

They also have totally different languages (in fact there are over 580 different Aboriginal languages), and different art styles. For example, consider the difference between the dot painting style of the Centre with the x-ray style of Arnhem Land.

It is probable that the diversity of Aboriginal people in Australia is greater than the diversity of people who lived in Europe in the 18th century. The significance of this diversity is expanded on in the section on subsistence hunting and foraging.

Aboriginal people have strong social links and cultural expectations which can result in social and cultural pressures different to those experienced by other sections of Australia's community. Aboriginal people both within and between groups, clans etc. will not always express the same view.

Within Aboriginal society, decision-making is a complex of: democracy; deference to age and standing; and strength or forcefulness. Thus different groups may reach very different decisions. An example of this diversity occurs in the uranium province of Western Arnhem Land where there are clan groups who actively support the conduct of further uranium mining and others who are implacably opposed to it. This is a feature just as much of non-Aboriginal society.

In summary, while Aboriginal society has many fundamental similarities it is not possible, and is often likely to be counter-productive, to encapsulate Aboriginal peoples views as homogeneous.

2.1 Historical context

It is likely that Aborigines had been living in Australia for at least 50 000 years before its colonisation by the British. Whilst we can never know exactly what the total population of the continent was in 1788, White and Mulvaney (1987), consider that "*an estimate of about 750 000 is a reasonable one*".

The population had fallen to less than 70 000 in the 1930s, after just 150 years of exposure to white 'civilisation' (Borrie *et al.* 1975,) by 1991, had increased to approximately 240 000 (Taylor 1993), which was still only 30% of pre-1788 levels (Figure 1).

Table 1 gives a chronological outline of the more significant events in recent Aboriginal history from the arrival of Captain Cook to the present.

2.2 Health

Pre-colonial Aborigines were a healthy vigorous people, with each family raising few, but robust, children.

Records written by European settlers and explorers often state that, upon first encounter, the Aborigines appeared to be in good health and free from disease. In general, their environment provided the resources for a nutritious diet.

The early colonists also believed that the Aborigines were nomads who roamed at random. This was a false picture, for although they might have moved camp often, each group hunted and gathered within a set area, usually returning to a particular place at the same season each year.

Once it was found that sheep and cattle flourished in Australia, conflict between the two peoples was exacerbated and hard to contain. Colonists regarded the land as an economic asset to be developed for their

Table 1: Chronology of significant events in recent Aboriginal history.	
1788 -	Captain Cook claims Australia as *terra nullius*.
1804 -	Settlers authorized by Lt. Moore to shoot about 50 Aborigines at Risdon Cove.
1830 -	Gov. Arthur tried to drive all remaining Aborigines on to the Tasman Peninsula. He failed.
1835 -	John Batman signed a treaty to 'buy' 243 000 ha of land in return for 20 pairs of blankets, 30 tomahawks, various articles and a yearly tribute.
1860 -	Victorian Board for the Protection of Aborigines set aside 1000 ha for all Aborigines.
1876 -	May 8. Truganini, the last full-blooded Tas. Aborigine died in Hobart.
1901 -	Jan.1 The Commonwealth Constitution stated that '*in reckoning the numbers of the people....Aboriginal natives will not be counted*'.
1921 -	The Aboriginal population was believed to have been at it's lowest point.
1937 -	Meeting of State and Commonwealth Ministers to resolve the 'Aboriginal problem' by separating them into 'uncivilized full bloods', 'semi-civilized full bloods' and 'half-castes' and placing them on 'small local reserves'.
1945 -	The first successful Aboriginal mining venture (to mine wolfram) was formed.
1946 -	Pilbara station walkoffs due to severe conditions.
1963 -	Charles Perkins staged the 'Freedom Ride' through NSW to attempt to end racial segregation.
1966 -	Gurindji walkoff from Wave Hill to their traditional country at Wattie Creek.
1967 -	May 27. The Australian people approved the ending of constitutional discrimination against Aborigines by a record 90.8% positive referendum vote.
1968 -	Yirrkala bark petition against the Nabalco mine.
1972 -	Aboriginal tent embassy set up outside Parliament House in Canberra.
1973 -	Recognition of Aboriginal rights by the Whitlam Government.
1975 -	Racial Discrimination Act came into force.
1976 -	Proclamation of the *Aboriginal Land Rights (NT) Act 1976*.
1978 -	Pat O'Shane became the first Aboriginal barrister. Ranger Uranium mining agreement signed.
1981 -	Pitjantjatjara people in SA were granted land rights.
1985 -	Canon Malcolm became Australia's first Aboriginal bishop.
1986 -	Ernie Bridge became the first Aboriginal cabinet minister.
1987 -	Royal Commission into black deaths in custody established.
1993 -	*Native Title Act* 'Mabo' proclaimed.

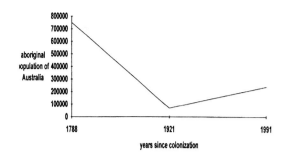

Figure 1. Graph of estimates of Aboriginal population from the time of European colonization (after White and Mulvaney 1987).

and leprosy amongst many) resulted in a rapid decline in the health of Aboriginal communities.

Reserves and missions

Alarmed by reports of massacres and the gross maltreatment of Aborigines, the British government brought pressure to bear on colonial administrators to protect the indigenous people. The response of the colonial administrators was to set aside reserves which were mostly quite small.

With the change to settlement life, major alterations in food habits were introduced. One reason for herding Aborigines onto settlements and reserves was to prevent hunting and gathering on land that had been taken over for stock-raising and cropping. This meant the abandonment of the traditional healthy mixed diet and the substitution of rations, which needed to be cheap, portable and non-perishable e.g. flour, tea and sugar.

Within the central and northern reserves, the missions and government settlements proved disastrous for the health of the Aboriginal inhabitants. Living habits suitable for a handful of people who moved camp regularly were totally unsuitable for several hundred settled permanently in one area. Authorities were slow to provide such basic facilities as piped water and sanitation.

2.3 Aborigines on cattle stations

On each of the large cattle stations of the north, a short distance from the homestead was the 'blacks' camp, where the remnants of the traditional owners of the land lived. The adult able-bodied men and some of the women were employed (but not necessarily paid) as labourers and stockriders, while other women worked as domestic servants in the homestead. Most pastoralists showed little concern for the health of their Aboriginal workers.

It is not surprising that in 1946 hundreds of Aboriginal employees and their families walked off pastoral stations in the Pilbara district of Western Australia, nor

economic gain, but to Aborigines it was sacred, a part of themselves. As Aborigines did not till the soil or live in settled villages, the intruders felt justified in moving them from one area to another.

This restriction of Aboriginal people to reserves, the introduction of diseases for which Aboriginal people had no resistance (e.g. typhoid, small pox, influenza

is it surprising that nearly 20 years later the occupants of the 'blacks' camp deserted Wave Hill Station in the Northern Territory and moved to Wattie Creek, to which they demanded rights of ownership.

Assimilation

By 1937 the concept of 'smoothing the pillow' was government policy and pervaded white society. The assumption was that Aboriginal health didn't matter because they were dying out anyway. The kind thing was just to provide a measure of comfort and dignity while this process took place.

This concept was translated into the adoption of the official assimilation policy which said that part-Aborigines were to be assimilated into the white community (regardless of their wishes), detribalised Aborigines were to receive an education, and the remainder were to stay on their reserves.

It was not until 1967 that authorities announced improvements were to be made in social welfare, health and housing and promised to acknowledge *'the very great diversity in the situations and needs of the Aboriginal and part-Aboriginal people throughout Australia'*. In other words, this was the end of assimilation as government policy.

3. ABORIGINAL ACTION

Since the mid-1960s, the most important gains in Aboriginal political, social and economic status — with consequent improvements in health, have resulted from the actions of Aboriginal people themselves (Lippmann 1981). The demand for land rights began to be heard throughout the country in 1966, when the Gurindji people walked off Wave Hill Station and demanded their rights to land.

In 1968, the Yirrkala people of Arnhem Land began a case against Nabalco and the Commonwealth with the aim of gaining recognition of their title to land under Australian law.

In January 1972, an Aboriginal tent embassy was set up on the lawns of Parliament House, Canberra, and this was backed throughout Australia by marches and demonstrations demanding land rights. Not only did these actions raise the consciousness of many white Australians but they also attracted world-wide media attention. Internal and external pressures on the Australian government have continued to be fully exploited by Aboriginal activists.

3.1 Traditional Aboriginal use of the savannas

For Aborigines, the land traditionally provided not only their daily economic sustenance but also the source of their origins and spirituality.

Three important elements of this traditional use are:

(i) foraging and hunting for game;

(ii) complex systems of inheritance of land responsibility which are interwoven with decisions over its use;

(iii) strong convictions that people should live on or close to important places in their ancestral country so that they can care for them properly.

These processes and beliefs have frequently been misunderstood by non-Aboriginal land users.

3.2 Subsistence Hunting and Foraging

Subsistence hunting and foraging not only results in Aboriginal groups ranging over country under their own control but also, as they are legally entitled to do, over country held by others.

While the general patterns of Aboriginal land responsibilities are similar throughout Australia, there are some regional differences which affect access to land. These differences may well influence decisions on land management and use. In the arid areas, it may be easier for groups to agree on what they want to do, because they are all seen to have an equal right to speak, whilst in the tropical north the possibilities for conflict between individual owners are perhaps greater.

People's conviction that they should be living near their ancestral country has more recently led to a population dispersal from larger centres of population, with small extended family groups living in isolated outstations and homeland centres.

Contemporary foraging and hunting for subsistence purposes now combines elements of traditional lifestyles with other elements introduced from non-Aboriginal society. Aboriginal people now hunt for both native and introduced flora and fauna and utilize European technology, notably vehicles, outboards and guns.

The incorporation of new technology and the use of cash in contemporary subsistence has not, however, destroyed its cultural significance. Foraging and hunting still provide people with satisfaction which is only partly related to the nutritional content of the foods obtained. It allows them to express their environmental knowledge stretching back over many generations, and continually reinforces their beliefs in the spiritual value of such knowledge; it is also an important medium of education, whereby both spiritual and ecological knowledge is handed on to succeeding generations.

Aspects such as these mean that subsistence activities cannot easily occur within land circumscribed by artificial boundaries such as those delineating leases, and that flexibility in access to resources is essential.

A common perception, is that subsistence is purely 'traditional', i.e. independent of the cash economy and that it is not worthy of recognition as a legitimate form of land use and management. A major reason for this is probably that the contribution of subsistence foraging and hunting to overall sustenance is extremely difficult to assess, and hence arguments in favour of its monetary support are hard to justify. (Young et al. 1991).

3.3 The Contemporary Contribution of Traditional Subsistence

The contribution of traditional subsistence to Aboriginal land management should be considered from two main perspectives; first, its contribution to Aboriginal sustenance, as the major non-monetary component in their economy; and second, its contribution as a deterrent to further land degradation.

Poor soil and seasonal variations in rainfall generally mean that the subsistence resource base in coastal northern areas has been little affected by pastoralism and is richer than that in the arid and semi-arid areas. Altman's study of Momega outstation near Maningrida (Altman 1987) showed that on average 47% of energy consumption and over 80% of protein came from the traditional subsistence component in the diet (Figure 2).

Meehan's study (1982) at another Maningrida outstation, the coastal community of Kopanga, complements that of Altman because it also takes marine resources into account. She found that marine resources, largely shellfish and fish, contributed 27%–37% of energy and 52%–75% of protein at different seasons of the year.

Traditional subsistence does not play such a significant role in the sustenance of Aborigines in all northern communities, particularly those which are larger in size and where the land resources have been significantly affected by non-Aboriginal settlement.

Failure to recognise the importance of traditional subsistence as a form of land resource use reflects the over-riding preoccupation with commercially productive forms of land management. The imputed monetary value of traditional subsistence production can vary considerably depending on how traditional subsistence is defined and on the costs and prices used.

Using market replacement values for proxy market foods, Altman (1987) estimated that bush food production was worth $1411 per capita per year (which equates to $3772 in 1994 terms). In 1979–80, this equated to 64% of total cash and imputed income at Momega outstation.

The existence of productive economies at outstations reduces external dependence on welfare and increases income status. The reduced dependency is due to a reliance at outstations on subsistence production which is estimated to account for between 48–64% of total income, compared with 5–10% of incomes at Aboriginal townships (Fisk 1985, Altman 1987).

These data suggest that decentralisation to outstations provides Aboriginal people with a means to economic advancement if the subsistence base and other income

Figure 2. Proportion of dietary energy gained from traditional subsistence at Momega Outstation near Maningrida (after Altman 1987).

sources such as artefact manufacture are well-developed and facilitated.

3.4 The outstation movement

The Blanchard Homelands Report (1987) defines outstations as *'small decentralized communities of close kin established by the movement of Aboriginal people to land of social cultural and economic significance to them. Outstations can also be referred to as 'homelands' and 'homeland centres'*.

Initially, outstations were established by groups decentralizing onto Aboriginal reserve land. More recently, groups have decentralized into National Parks (Kakadu and Cobourg) and onto Aboriginal and non-Aboriginal pastoral stations.

The Blanchard Homelands Report (1987), delineated 588 homeland communities with an estimated population of 9538 people and 111 excision communities with a population of 3921 people. Altman and Taylor (1987) estimated that in 1986 approximately 17 500 Aboriginal people were living at about 500 outstation locations. This represented about 8% of the total Australian Aboriginal population.

The environments in which outstations are located vary significantly. In the Northern Territory, outstations tend to be more numerous in the monsoonal areas of the Top End. Outstations are also located in similar environments in the north of Queensland and in the Kimberley region of Western Australia. They are also scattered throughout the inland savannas of the south Kimberleys, the Pilbara and the Northern Territory and in arid regions in the south of the Northern Territory, the north of South Australia and the east of Western Australia.

The environmental differences have important implications for the variability of outstation economics. There are also locational differences that influence both formal employment and market exchange activities at outstations. For example, some outstations are located at remote locations with extremely limited market linkages; others are located near townships with employment opportunities.

The initial shift to outstations occurred in the period 1972–1976. Outstations were originally reoccupied for a complex set of reasons that included cultural (desire to protect sacred sites), social (need to reinforce traditional family structures) and political (desire for autonomy) factors.

The second phase of the movement dates from the late 1970s when it became increasingly recognised in the wider community that structural unemployment was a longer term reality for all Australians. The payments of unemployment benefits changed the economic rationale for decentralisation because the economic benefits of outstation living increased markedly.

The Community Development Employment Programme (CDEP) is a method of people working for the dole and receiving additional assistance from the Government to provide the resources to enable them to work. Resources such as materials, tools and training are provided by the programme to enable the CDEP group (which may be an entire outstation population, or part thereof, or even a group of outstations or communities) to carry out an employment programme. The introduction of CDEP has therefore enabled outstation groups to extend the value of their unemployment benefits as well as providing a mechanism for developing their surrounds.

4. ECONOMIC USE OF THE SAVANNAS

4.1 Aboriginal Economics

Today, only a small proportion of Aboriginal and Torres Strait Islander people follow their traditional subsistence patterns and most seek their livelihood through the wider economy. Despite their overall relative poverty, Aboriginal and Torres Strait Islander people can have a significant influence on both local and regional economies. For example, Aboriginal people contribute one-third of the central Australian economy through citizens' entitlements, grants, earned income and Aboriginal bureaucracy (Crough *et al.* 1989). Very few, however, are self-employed, in contrast to other Australians.

4.2 Employment and Income Patterns

According to the 1991 Census, the overall rate of unemployment among Aborigines was about 30.8%, about three times that of the general population.

Aboriginal people in the Northern Territory have average money incomes that are less than half those of other Australians, and also less than the incomes of Aboriginal and Torres Strait Islander people in other states. Although rural Aborigines have greater access to non-money income through their ability to hunt and gather bush food, bush food is now only a proportion of their total diet. Because of limited employment opportunities for Aboriginal and Torres Strait Islander people living in rural communities, their participation rates and employment rates are even lower than the overall averages for Aboriginal and Torres Strait Islander people.

Aboriginal people in part of the Northern Territory have access to income linked to mining royalties. For some communities income from selling art and artefacts and through tourist developments is increasing. However, in the majority of rural Aboriginal and Torres Strait Islander communities, full-time employment is available to only a small minority working for the council, school or health centre or store.

More employment opportunities, particularly part-time ones, are now being developed through the CDEP scheme, administered by community councils.

The private sector is also contributing to expanded employment opportunities for Aboriginal people. Some Australian mining companies have developed Aboriginal employment schemes. A number of studies however, have questioned the prospects of Aboriginal and Torres Strait Islander people achieving greater employment in the mining industry (Altman 1983). These have shown that in the last ten years the proportion of Aboriginal people employed in the industry has remained very small. Indeed, recent research has shown that the representation of Aboriginal people in mining employment has actually declined.

Aboriginal people suffer from very low social and economic status in Australian society. This is reflected in measures of education, employment, income and housing. Lack of appropriate education and training continues to be a powerful barrier to improving their economic prospects.

5. IMPACT ON THE SAVANNAS' ECONOMY

Australia is coming out of a severe recession, unemployment levels are still high, foreign debt is high, and business confidence is only starting to rise.

As is usual during a period of economic and social difficulty, large sections of the community are desperate to blame someone else for their problems.

Sometimes it is newly-arrived migrants, who are supposed to be taking jobs away from 'Australians'. At other times it is environmentalists, who are said to be undermining 'development' because of their insistence that decision-makers take more account of environmental factors. This might be acceptable in a boom, but, so the argument goes, the country cannot afford such a luxury at this time.

Another group which is more subtly being blamed is Aboriginal people who own land. Many Australians are not prepared to openly criticise traditional Aboriginal land owners, although many are quite happy to blame their 'white advisers'. The *Aboriginal Land Rights (NT)*

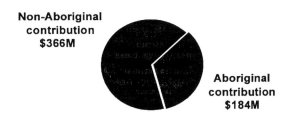

Figure 3. Contribution to the economy of central Australia by Aboriginal people (after Crough *et al.* 1989).

Act 1976 (ALRA) has been a particular focus of criticism for a number of years, but even where Aboriginal people have no real rights to control development, particularly in Western Australia and Queensland, Aboriginal people and their advisers have still been subjected to strong criticism.

Aboriginal people have made important gains in the Northern Territory. But there is no doubt that certain industry groups believe that Aboriginal people have achieved too much, and these groups are prepared to voice their concern very publicly. The argument is that the costs of the recession have to be borne equally by everyone in the community. In a depressed economic climate, these views appear perfectly reasonable. They are very much like the nineteenth century view that the rights of Aboriginal people could be ignored because economic growth, industrial development and 'progress' were more important. In the end Aboriginal people would benefit once they became assimilated into the non-Aboriginal population.

The reality is, however, that Aboriginal people are, by themselves, significant contributors to the economic welfare of the Savannas, the region of Australia in which they have their largest representation (Crough *et al.* 1989). These authors demonstrated that central Australian Aborigines contributed at least $184 million to the central Australian region in 1987–88. These statistics destroy the myth that Aborigines have only marginal significance to the local economy (Figure 3).

While similar studies are needed for northern Australia, it is useful to examine the results of this study of the central Australian region. We think the general patterns will be similar.

Most importantly, the study tells us that the Aboriginal sector cannot be regarded merely as a welfare transfer from the government. Together, these monies directly keep one-third of the Central Australian economy afloat. The indirect effects of these expenditures, and the extent to which Aborigines comprise an intangible asset for the region (as in their cultural importance for the tourist industry) enlarge this role further (Crough *et al.* 1989).

Whilst it is not possible to directly extrapolate the experience of Central Australia to the whole of the savannas, the results of Crough *et al's*. work demonstrates that Aboriginal people in many ways are significant contributors to the economy of the regions in which they live and not just by way of grants from government by virtue of their Aboriginality.

5.1 Constraints

There are three major constraints to Aboriginal economic development:

(i) the characteristics of the Aboriginal population;

(ii) the political environment;

(iii) the size and state of the local economy.

The characteristics of the Aboriginal population of the savannas have been dealt with previously, their employment, health, education and place of residence are all significant factors.

The political environment within which Aboriginal people live presents problems. Conservative governments in WA, NT and QLD at various times have constrained Aboriginal development by opposition in many forms, the often patronizing approach of service delivery and the trend to funding cuts and mainstreaming of Aboriginal programs are all constraints at one level or another.

The local economy is a constraint for Aborigines because its narrow base limits the options for Aboriginal enterprises. Furthermore, the relatively low rate of regional growth at present and the tendency for many industrial sectors in the savannas to be controlled by well established firms makes market entry difficult but not impossible, as is demonstrated in more recent times by significant Aboriginal business ventures in the NT.

5.2 Conflicts and development in the savannas by industry

Throughout this paper it will have become abundantly clear that conflict at one level or another insinuates itself throughout Aboriginal involvement in the savannas; and yet as has also been made clear, Aboriginal people contribute enormously to the savanna region, economically, socially and culturally. This section looks in a little more detail at each of the main industries of the savannas.

5.3 Pastoralism

The pattern of land tenure in the savannas in the 1990s is the result of more than a century of non-Aboriginal settlement. In this section the discussion is focused on the NT.

By 1876, only a handful of pastoral leases had been granted, all of which were in central Australia. It appears that the first pastoral station stocked in the Top End was in 1879 (Reid 1990).

By 1881–82, however, virtually all of the Northern Territory was leased to 'pastoralists' or held under application. Only 33,000 of the 523 620 square miles of land area were not dedicated, on paper at least, to pastoral activities.

The land tenure maps certainly give the impression of significant economic development in the Northern Territory during this period. The problem was that most of the development was little more than some official in a government department drawing lines on a map. Most of the applications were speculative, and much of the country was, and still is, unsuitable for pastoral purposes. Much of the country had not even been explored by non-Aboriginal people.

Despite many leases being relinquished, there is no doubt that by the mid-1880s, the pastoral industry had finally gained a foothold in the Northern Territory. From this period onwards the disruption to Aboriginal society began to accelerate. This is evident from the reports of the Government Resident of the Northern Territory, which from 1884 begin to express considerable concern about Aboriginal-pastoralist relations. In his 1884 report, under the heading *Aboriginals and Settlement*, the Government Resident warned:

'*I fear unquiet times may be expected in connection with native tribes. The blacks are beginning to realise that the white man, with his herds, and his fences, and his preservation of water, is interfering with that they properly enough, from their point of view, regard as their natural rights. Their grounds and game reserves are being disturbed, and their food supply both diminished and rendered uncertain…The natives will resist the intrusion of the whites and regard themselves as robbed of their inheritance…That settlement and stocking must and will go on is certain - that outrages will be committed by both sides is probable; but even those who do not claim to be philanthropists are not satisfied with the contemplation that the blacks are to be improved off the face of the earth*'.

By the mid-1960s, the distribution of leases and their boundaries had essentially stabilised. The major exception to this, is of course, as a result of land claims under the ALRA, which have meant that much of the unalienated Crown land and the former reserve lands, has become Aboriginal land.

While the ALRA has enabled some Aboriginal landowners in the Northern Territory to gain title to former reserves and to claim unalienated Crown land, the legislation was not able to address the land needs of Aboriginal groups in pastoral areas. In the Kimberley and Cape York limited access to funding programs has frustrated Aboriginal people's aspirations.

As a result many Aboriginal people continue to live under appalling conditions without secure title to their traditional land and access to basic facilities. Over the past two decades the land needs of these people have been partially addressed through pastoral property purchases and in a much less satisfactory way, through the inadequate and ill-conceived pastoral excisions process.

Since 1974 the purchase of leases by Aboriginal traditional owners has been funded by various Commonwealth bodies including the former Aboriginal Land Fund Commission (ALFC), the former Aboriginal Development Corporation (ADC), and in more recent years, the Aboriginal Benefits Trust Account (ABTA) and the Aboriginal and Torres Strait Islander Commission (ATSIC). As a consequence, a total of 21 properties have been acquired by Aboriginal interests in the Northern Territory, 22 in the Kimberley region and 8 in north Queensland (ATSIC property register).

These properties were acquired in order to provide Aboriginal traditional owners with a secure land base to help them meet their social, cultural and economic needs. While pastoralism is obviously seen as the major economic activity, the marginal nature of the majority of these properties and their relatively poor condition at time of purchase has placed significant constraints upon future productivity. In addition, it is recognised that Aboriginal landowners have other land use needs and aspirations including traditional and subsistence use, community development, tourism, conservation and so on.

ATSIC has played a major role in providing funding to support property development and management and has worked closely with other Aboriginal organisations to expand the support services available to Aboriginal cattle enterprises.

Following the passage of the *Native Title Act 1993* in December 1993, the Commonwealth Government announced its intention to establish a National Aboriginal and Torres Strait Islander Land Fund specifically designed to purchase land for Aboriginal people unable to benefit from the Native Title legislation.

An important feature of this fund will be the recognition that support for ongoing management and development of Aboriginal properties is vital.

The controversy over pastoral lease purchases by Aboriginal interests was fuelled at the end of 1991 following the approval by the Minister for Aboriginal Affairs for the use of Aboriginal Benefits Trust Account funds to purchase a number of leases. The NT Government criticised the purchases, the land claims process, and the Land Councils and said that there was '*a loss of valuable land to our pastoral industry*'. They argued that Aboriginal-owned properties were generally stocked well below their estimated carrying capacity, ignoring the fact that most of the leases when purchased from non-Aboriginal interests had either no cattle or relatively low numbers of cattle on them.

The pastoral industry has a long history of Aboriginal involvement although this has been generally at a non-managerial level. Whilst Aboriginal pastoral properties have produced mixed results it is important to recognize that this is largely the result of inexperience, lack of access to capital and unreasonably high expectations. Although the current growth prospects for pastoralism in general are not high, the industry can play an important role for individual communities.

In order to understand the prospects and economic potential for Aboriginal pastoral properties it is important to understand the scope of the industry in which they operate. Historically the pastoral industry has been seen as one of the mainstays of the Northern Territory economy. A closer examination however shows that the pastoral industry is no longer a particularly large sector of the NT economy nor does it offer great opportunities for profitable activity.

There are currently 269 pastoral properties in the Territory. Of these, 67% are privately owned by non-Aboriginals, 14% are owned by Aboriginal interests and the remaining 19% are owned by companies. In recent years the proportion of buffalo and cattle enterprises owned by foreign interests has increased. Small scale or family concerns have generally not been able to raise enough capital for development.

Cattle and buffalo production during 1991/92 in the NT was valued at $120 million which is about 2.9% of GSP (Gross State Production). This figure is put further into perspective when we realise that the total Northern Territory GSP is just 1.1% of Australia's GDP so what is tiny in NT terms is even smaller in national terms i.e. merely 2% of 1.1%. Employment in the industry accounts for less than 2% of the NT workforce.

Given this industry context, it is clear that Aboriginal pastoralists, under the physical, educational, economic and environmental constraints described earlier, will, in most cases, find it very difficult to run cattle enterprises which can be considered to be successful on purely economic criteria. By and large they enter the industry with considerable limitations in the form of marginal properties with diminished natural resources, declining infrastructure and lack of their own capital to invest except through assistance from funding agencies.

The indisputable fact however, is that the gains in social and cultural fields which have been achieved by Aboriginal people owning their own land, are of great significance. Economic self-sufficiency may be achievable in the long term but at this stage it is too early to tell. Instead of seeing Aboriginal cattle properties purely in pastoral production terms, Aboriginal people prefer to see them as examples of integrated rural development.

5.4 Mining

Of all the industries in which conflict occurs with Aboriginal people, exploration and mining would have to be the most bitter. From its earliest days the industry has a history of taking little account of Aboriginal views.

In 1928 J.W. Bleakley, the Chief Protector of Aboriginals in Queensland, recommended to the Commonwealth Government that there was a need to extend some of the existing reserves and establish others in areas such as Arnhem Land, the Tanami region the Petermann Ranges and Melville Island. By the time the first conference of Commonwealth and State Aboriginal authorities was convened in 1937, mining on those reserves had already become a contentious issue.

Subsequently the lives of people living on reserves was continually subject to change through boundary changes, revocation (in part or *in toto*) and people were forcibly moved largely as a result of the needs of miners.

In spite of this history, since the inception of the *Land Rights Act* in 1976, there have been nine mines established on Aboriginal land, two of which were discovered under negotiated exploration agreements. By 1994, Aboriginal people of the NT had entered into agreements with explorers covering more than 50 000 km^2 (an area one quarter the size of Victoria) with others in the pipeline over a further 32 500km^2.

In the case of two mining projects, Jabiluka and Koongarra, Aboriginal consent to mining on their land has been overridden by Commonwealth policies over uranium mining. Despite this, the attacks on Aboriginal rights by both the industry and conservative governments have been virulent, expensive and extensive. Often times in the debate, truth has been a major casualty.

Aboriginal people are not necessarily anti-development or anti-mining or pro-mining. Interestingly, Aboriginal people have historically been involved in the mining industry for untold generations. There are a number of examples of Aboriginal extraction sites around the savannas. At these sites Aboriginal people extracted resources of use to themselves such as ochre, pipeclay and stone. Most mines were open cut, but there are examples of ochre mines with underground operations. Most appear to have ceased operations no later than 50 years ago, but at least one ochre mine in the NT is still functioning (Horton 1994).

Whilst virtually all Aboriginal land in the savannas has previously been explored, it is recognised that past mining and exploration is a reflection of the existing technology, knowledge and commodity prices. All of these have changed enormously and there is no doubt that some areas of Aboriginal land are now highly prospective.

Many Aboriginal people see the industry as their only way to achieve self-sufficiency and independence of the social welfare, whilst others of course are totally opposed to their country being mined. Mining is of interest to Aborigines because of the opportunities it presents for employment and income, for example through the payment of negotiated royalties and royalty equivalents through ABTA (as occurs in the NT).

All of the major mining projects in the NT are located on Aboriginal land. In 1991 these mines produced over AUD$1 billion (80%) of the total NT mineral productions. In addition, the gas and oil fields added over AUD$15 million and in 1993 exploration expenditure on Aboriginal land alone was in excess of AUD$20 million.

The Australian Mining Industry Council (AMIC), the industry's peak body, has argued in many forums, that:

- nothing should be done that might stifle discovery and development;
- the mineral wealth belongs to the whole community, and no landowner should be in a position to 'lock away' such valuable resources;
- any form of assistance which would significantly set Aboriginal people apart from the rest of the community would be disruptive;
- traditionally, the minerals and metals of modern civilisation had never been part of Aboriginal life;
- governments must retain the power to grant exploration and mining titles over areas independent of other land use;
- in granting land rights to Aborigines the land titles should not carry with them any measure of mineral ownership;
- mining will only occupy a minute proportion of the areas presently reserved for Aborigines;
- the rights of Aborigines in relation to mineral developments on their land should take place within the existing legal framework, and;
- it would be undesirable to allow negotiations for an interest based on the value of the minerals.

After considering the industry's position in the *Second Report of the Aboriginal Land Rights Commission*, Justice Woodward (1974) concluded: '*I believe that to deny to Aborigines the right to prevent mining on their land is to deny the reality of land rights*'.

When the Commonwealth announced in 1989 that the Industry Commission would undertake an inquiry into mining and minerals processing, many in the industry and government expected that the Commission would recommend substantial changes to the Act. While the Industry Commission did recommend a number of changes to the ALRA, it generally reinforced the views of Justice Woodward. On the issue of whether the Act was undermining the industry in the Northern Territory, the Industry Commission (1991) concluded:

'*The holding of land rights may lead to smaller levels of mining (and more particularly exploration) activity in the Territory relative to those which would have occurred. However, provided Aboriginal landowners face appropriate incentives, it would be wrong to conclude from this that land and sub-surface resources were not being devoted to their socially optimal use*'.

5.5 Tourism

Tourism is the industry most regularly cited as an area for greater Aboriginal participation. The pace of growth in tourism remains rapid and there is little doubt that many tourists have at least a cursory interest in Aboriginal culture and lifestyles. Along with flora, fauna and Australian lifestyle, Aboriginal culture is an important part of the drawcard Australia has to offer the tourist (PATA 1992).

Aboriginal participation in tourism can take many forms and it is crucial that their organisations and communities understand these options before committing themselves to any involvement. Most importantly, Aborigines must develop structures to control tourism more effectively, so that returns to Aborigines through arts and crafts sales and other enterprises are maximised.

While difficult to definitively value, the gross worth of tourism development on Aboriginal land in the NT is well in excess of AUD$15 million p.a.

For example, Aboriginal landowners own the two major hotels in Kakadu, the Gagadju Crocodile and Gagadju Cooinda, along with the Kings Canyon Resort in Central Australia.

In the Top End, 14 separate companies operate specialist tours under agreement with Aboriginal owners for such things as hunting and photographic safaris, scientific expeditions and Aboriginal cultural experiences. Between them these tours alone generated approximately AUD$0.6 million in 1992.

6. NATIONAL PARKS/ PROTECTED AREAS

Since the late 1970s, involvement in national park management has become one way for Aboriginal people to regain some measure of control over their traditional lands. It has also meant access to training and employment (e.g. as rangers) and to resources to manage their lands.

For conservation agencies, Aboriginal involvement provides a cultural dimension which results in better management and an enhanced experience for visitors. The disadvantages for Aboriginal people include the necessity to share decision-making with a conservation agency, and to share their country with visitors (Horton 1994).

The only Aboriginal-owned national parks are in the NT, and they include Kakadu, Uluru, Nitmiluk and Gurig. In Queensland, the *Aboriginal Land Act 1991* allows ownership if a claim is successful. In NSW and WA provisions are being made for traditional owners to be involved in management but in the southern states no such arrangement exists.

Kakadu National park was one of the first areas to gain World Heritage status for both biological and cultural criteria. While it only became a national park in 1979, Kakadu has become an Australian icon. National policy on many conservation and Aboriginal issues have been forged and continue to evolve there.

The greatest achievement of the park lies in the area of joint management with Aboriginal people. In particular, the Board of Management has 10 of 14 members who are traditional owners of the region. Additionally, there is a growing proportion of Aboriginal staff employed in the park.

Kakadu is the prime national and international drawcard for the tourism industry of north-western Australia, most other tourist attractions do not have the profile of Kakadu and, irrespective of their own inherent value, piggyback on the drawing power of Kakadu.

As for tourism, experiencing Aboriginal culture and history are some of the most common reasons for people to visit the park (Braithwaite 1994). Unfortunately, the aspirations of Aboriginal people and the requirements of western conservation and tourism do not always perfectly coincide and compromises are necessary (Braithwaite 1992).

It is obviously even more disappointing for Aboriginal people in those areas of the savannas where they are not able to have a meaningful say in the management of parks. It can only be hoped that governments will recognise the advantages of joint management with Aboriginal people and move in this direction.

6.1 Transport and Communications

Traditionally remote Aboriginal communities have relied almost totally on non-Aboriginal owned transport companies including air, road and barge firms. This has meant that people are totally at the mercy of these companies for both the level and cost of service, and where monopolies exist it is fair to say that companies have taken advantage of the fact.

More recently however, transport and communications has been a growth area for Aboriginal enterprises, these being used to provide services to remote communities and protect Aboriginal culture. The operations of Aboriginal airlines, fuel franchises and the more recent barging joint venture in the Gulf of Carpentaria are examples of Aborigines taking greater involvement of important modes.

The entry of Central Australian Aboriginal Media Association (CAAMA) into radio and television through Imparja is a further example of providing services to people on a commercial basis. With the expansion of satellite communications there are more opportunities for successful Aboriginal involvement in this area.

6.2 Construction and Retailing

The construction industry has owed much of its historical growth to Aboriginal projects. Yet direct Aboriginal involvement in construction is small. In more recent times some of the Outstation Resource Agencies have purchased mud brick making machines and are using them to construct more comfortable and appropriate dwellings at outstations in the NT but the majority of the construction work on communities and outstations is still carried out by non-Aboriginal people.

Likewise, retailers generally have benefited greatly from Aboriginal demand, yet Aborigines have not fully mustered their consumer strength to date. They are starting however; examples include the Aboriginal owned stores on smaller communities and in central Australia, and Aboriginal interests in larger shopping centres.

6.3 Military Activities

Two hundred and eighty nine Aboriginal men from NSW, Vic and Qld went overseas with the 1st Australian Infantry Force in World War I. Of these, 44 died and another 59 were wounded. At the outbreak of WWII many Aboriginal men again volunteered for overseas service. In May 1940 the Military Board advised recruiting centres that the recruitment of non-whites was 'not desirable'. Adverse publicity followed and by 1942 an estimated 3000 had enlisted. A further 3000 worked as civilian labourers and another 200 worked in patrol and coast-watching duties.

The war's end brought disillusionment for many as they returned to civilian restrictions unlike the equality they had experienced in the services. It was only many years later that Governments finally acknowledged their contribution.

More recent involvement with the military has come with the conduct of the Kangaroo exercises in the north, where the Army in particular has sought access to extensive areas of Aboriginal land. Generally the relationship between the military and Aboriginal traditional owners has been sound. The military maintains a public relations unit who try to ensure that all appropriate clearances are in place before the exercise commences.

Unfortunately, there have been some instances of site desecration, some examples of excessive damage to country and at least one example of the transport of feral animals from interstate (in this instance, a canetoad). These incidents can usually be ascribed to individuals either ignoring instructions or misunderstandings. Regardless of the reason, these indiscretions do create an image in peoples' minds that make them wary of on-going involvement with the military.

6.4 Cropping, Horticulture and Forestry

Small-scale cropping and horticulture has been an area in which some Aboriginal people have expressed an interest. For example, from the early mission days, market gardens were planted extensively. These subsequently fell into disrepair due principally to the distance from markets. These days they are largely only used for subsistence food supplies.

Forestry has been experimented with on Aboriginal land since the 1960s, where a number of species of timber including cypress pine were planted in areas such as Maningrida, Cobourg Peninsula and the Tiwi Islands. Most of these plantations failed due to climatic conditions and, again, the distance from markets. Today the only surviving plantations are those on the Tiwi Islands.

6.5 Arts and Crafts

Art and artefact production are an important source of income for Aboriginal people. Aboriginal arts are manysided and interconnected. In pre-contact society visual arts were associated with ceremonies and served educational, religious and social purposes. There was no distinction between 'fine art', 'decorative art' and 'craft' such as occurs in Western art.

Aboriginal artforms include sculpture, engraving, painting on the body, ground, rock or bark, and weaving mats and baskets. The object was, and still is, less important than the process of producing it, the ritual purpose it may serve and the story it tells, all of which are inextricably linked to land, religion, ethics and social structure.

Bark painting was the first form of Aboriginal art to be affected by the European desire to collect portable art objects. From the late 1920s regular art production was sponsored by missionaries in Arnhem land, with the aim of achieving economic self-sufficiency and reinforcing Aboriginal identity.

Subsequently, Aboriginal art has evolved through watercolours, acrylics and oils. Additionally, both men and women have been producing items that were originally of customary use, for sale, (e.g. coolamons, baskets, mats, musical instruments and weapons).

Art and artefact production has significant economic importance as it is one of the few sources of non-government cash and has the advantage of earning money for activities connected with Aboriginal identity and tradition. There are many problems associated with their sale though, and these include:

- the price is usually determined by outsiders, not by the producers, and in the art market as a whole, much of the profit goes to agents and galleries;
- the lack of Aboriginal control of the industry, although the formation of the Association of Northern and Central Australian Aboriginal Artists Agency (ANCAAA) has assisted to regain some of that control;
- the imitation of Aboriginal images by governments, commercial manufacturers and non-Aboriginal artists (for example, in 1988 the new plastic AUD$10 note used a morning star pole by Terry Yumbulul without permission).

6.6 Conflict resolution mechanisms

The resolution of conflict between Aboriginal and non-Aboriginal people in the savannas is not easy and cannot occur until there is both the will and the means. Aboriginal people argue that the means have always been there, that is, the recognition of their rightful place in this country as the traditional landowners. They also argue that despite the enormous injustices inflicted upon them, such as dispossession and genocide, they have always demonstrated a will to endeavour to live in harmony with the non-Aboriginal population.

From the perspective of many of the non-Aboriginal population of the savannas there are a number of areas where progress can be made. This section considers some of these.

6.7 Documentation of traditional knowledge

To date, the documentation of traditional Aboriginal knowledge has primarily been the domain of anthropologists. A strong focus has been placed on describing kin systems, myths, systems of classification etc. The emphasis has been on Aboriginal culture rather than Aboriginal knowledge *per se*.

Anthropologists have sometimes lamented the lack of involvement of zoologists and botanists in studying Aboriginal knowledge (Berndt 1989). There has not been a tradition of study in these areas from those disciplines. In this era of growing interdisciplinary study, we are likely to see however a greater interest in such collaboration as evidenced by the production of publications such as the Aboriginal Pharmacopoeia jointly by Aboriginal people and the Conservation Commission of the Northern Territory. It is important however, that matters such as the intellectual property rights are also considered seriously and as a legitimate right.

6.8 Helping European science

A recent well-documented example of pooling of knowledge is that of a collaboration between the Mutijulu community, CSIRO scientists and ANCA staff in the survey of the fauna of Uluru National Park (Reid et al. 1992). An historical context on the former and present distribution of species can be provided by Aboriginal people as can details of the natural history of species of plants and animals. Looking at a problem from a different cultural context can enrich our understanding.

Thus it is our view, that Aboriginal people can contribute to helping European science. This matter is expanded on further in the section on research on Aboriginal land.

6.9 Helping European management

European land managers can learn a great deal from Aboriginal people. They have a continuous history of sustainable utilisation of Australia's land and marine resources stretching back at least 50 000 years. By contrast, since European settlement began, 20 species of mammals, 22 other vertebrate animals and 83 vascular plants have become extinct.

Estimates of the extent of land degradation vary, but figures like 60% of utilized rangelands requiring some type of rehabilitation are not uncommon (Pearce 1994).

One area in particular, fire management, is widely touted as an area where Aboriginal knowledge is valuable for contemporary management. While this is valid, the application to a modern situation must be done thoughtfully (Braithwaite 1992). It is our view therefore, that considering Aboriginal views and knowledge can significantly improve the present management regimes.

6.10 Appropriate technology

Culturally appropriate tools for enhancing the quality of existence for Aboriginal people is an area of research which has received some attention (Foran and Walker 1986). Such appropriate technology may result in more efficient use of financial/social service resources of the Australian community.

Organisations such as the Centre for Appropriate Technology in Alice Springs, have a great deal to offer. The development of solar power generation equipment and other simplified machine technology and their longevity in remote communities are evidence of their successful approach.

6.11 Exotic plants and animals

An invasion of exotic plants and animals followed the arrival of Europeans two centuries ago. They are clearly a great threat to the distinctly Australian ecosystems characterised by high levels of endemism and adaptation to low levels of nutrients and erratic rainfall (Braithwaite 1991). Traditional Aboriginal knowledge accumulated gradually over millennia does not deal well with the very recent problems caused by these exotic organisms. Western science on the other hand has developed methods of control and indeed continues to develop new methods in this area. Consequently this is an area of scientific knowledge particularly valued by Aboriginal people.

It also critical that governments recognise not only the science but also the cost associated with their control. Invariably these problems are the result of non-Aboriginal mismanagement not Aboriginal mismanagement and Aboriginal people do not have the physical, technical or, in particular, financial capacity to deal with them. For example, between 1991 and 1996 it is expected that in excess of $8.5 million will have been spent on the control of one 7000ha infestation of *Mimosa pigra* in western Arnhem Land, clearly the Aboriginal landowners of the region do not have access to this level of funding. The obligation to meet such costs must rest with governments to ensure that many vital ecosystems are protected.

7. MEASURES OF THE HEALTH OF THE LAND

The concept of the health of the land is important because of the changes which have resulted from the exotic invasion mentioned above. Losses of desired traditional resources are certainly motivating these concerns as referred to in the above section on 'Helping European Management'. However, the changes caused are often outside the traditional signs and symptoms of good management of the land. Aboriginal people value the development of indicators which are good measures of the health of the land and suitable for use by Aboriginal people.

In a sense, this is a demand for broader indicators of sustainability more appropriate to an Aboriginal concept of land, rather than more narrowly based indicators such as those appropriate to cattle producers. Work currently being done on the development of a Rangelands Strategy by the Commonwealth Government and the Global Desertification Forum are likely to assist in the development of such indicators.

Additionally, the proposal currently being put to Government for the formation of a Co-operative Research Centre for the Tropical Savannas is one that has the full support of Aboriginal people. The proposed research program, will amongst other things, seek to develop indicators of economic sustainability appropriate to this region.

7.1 Winning Arguments with Bureaucrats and Politicians

There is evidence that the people with ultimate power in Canberra accept arguments based on science over

those based on Aboriginal knowledge/ rights. The findings of the Resources Assessment Commission (1991) on the proposal to mine Coronation Hill in Kakadu National Park was unusual in presenting both arguments as was the report of the Industry Commission (Industry Commission 1991). In these and other recent instances it may also be a reflection of the growing sophistication of Aboriginal people and their organisations in dealing with issues at a national level.

7.2 Research On Aboriginal Land

Ultimately, Western science and Aboriginal culture represent two very different systems of knowledge. Aboriginal knowledge is relatively slow to change and be learned, and is the source of status of elders. Science, on the other hand, is seen to offer great status in a relatively short time to young researchers who bear all the usual signs of the inexperience of youth. Aboriginal people, on the other hand, do not understand the full method of science with its publication and peer review allowing rapid access to vast international experience through the literature.

The notions of cause and effect are fundamentally different (Wellings, unpublished manuscript), one logic-based and the other myth-based. Each tends to believe in the exclusive validity of their culture-bound approach.

There is also a widespread notion among biological and physical scientists that studying on Aboriginal land is more trouble that it is worth. The additional layer of approval for work on Aboriginal land has acted as a deterrent. It is for this reason that Aboriginal land like Arnhem land is very poorly known by Western science.

Where Aboriginal knowledge is shared, there is often poor recognition of intellectual property rights by scientists. They tend not to acknowledge Aborigines as peers in the knowledge-gaining process. Aborigines are often only accorded the same status as machines in the process. This can result a sense of having been cheated or dispossessed of their culture.

The strong belief in the power of the scientific method by scientists is seen as arrogant and an obstruction to mutual respect and understanding by some Aboriginal people. Most scientists view science as a far superior method of finding out about the world. On the other hand, some Aboriginal people counter any examples purporting to show the value of science with one of the following arguments:

a. Let me know something I do not already know?
b. What good is that knowledge to Aboriginal people?
c. Where are the answers (to a particular problem)?

Aborigines prefer intuition and action to slow and deliberate logic, order and analysis (Wellings, unpublished manuscript). The time and discussion required to formulate intuitive action by a community is often seen as time wasting by scientists. The impatience and business of people of European culture whose prime interest or professional focus is not Aboriginal culture is also a hindrance to communication and understanding.

On the other hand, Aboriginal people see the meticulous collection of information by scientists long after the appropriate course of action has become clear to them as typical procrastination by the European culture. In essence, scientists assume you need to formulate the problem and collect the relevant data to solve a problem. Aboriginal people assume the data needed has already been collected and it is a matter of drawing on the life experiences of the whole community, particularly the elders.

Research can have a large role in building better relationships between Aboriginal and non-Aboriginal people. At a practical level, tips to make it work, or at least get off to a good start, are needed. The National Health and Medical Research Council endorsed the following guidelines (Maddocks 1992) for research among Aborigines and Torres Strait Islanders:

- appropriate consultation;
- the opportunity for informed consent;
- agreement that the proposed research is of benefit;
- that the work is of scientific merit;
- opportunities for Aboriginal participation;
- opportunity for regular feedback;
- opportunity for joint ownership; and
- agreement on control of information.

8. Conclusion

Aborigines have a substantial influence in some key components of the savanna's economy. However, its economy has a narrow base because isolation and high costs make many enterprises unviable. For remote Aborigines, the constraining influence of the local economy is even more harshly felt. There is a danger for Aborigines that, because opportunities are limited, a number of Aboriginal communities may enter the same type of enterprise and compete with each other. The result may not necessarily be for the benefit of Aborigines as a whole.

Obviously not all Aborigines or non-Aborigines have the same views. 'Redneck' views that Aboriginal knowledge is without value are as unhelpful as those 'do-gooders' who view Aboriginal people as 'ecologically noble savages' who have a mysterious knowledge of all ecological problems and always behave with great ecological prescience.

Effective communication leads to an understanding of the other viewpoints. This leads to mutual respect.

However, this can only happen when there is desire to do this. Reconciliation is dependent on the recognition that it is in the national/community interest.

Equally, the ecologically sustainable management of the savannas' natural resources, if it is to be successful, must involve Aboriginal people. To do this successfully, it must begin with policy making which addresses Aboriginal rights, not just Aboriginal needs.

'Indigenous people and their communities, and other local communities, have a vital role in environmental management and development because of their knowledge and traditional practices. States should recognise and duly support their identity, culture and interests and enable their effective participation in the achievement of sustainable development.' (Principle 22 of the Rio Declaration).

Acknowledgements

We would like to acknowledge the efforts of Robbie Braithwaite in searching out the information for the Chronology and Bronwyn Motlop for her patience and keyboard skills when faced with numerous versions of the paper.

References

Altman, J.C. (1983). *Aborigines and Mining Royalties in the Northern Territory.* AIATSIS, Canberra.

Altman, J.C. (1987). *Hunter-Gatherers Today.* AIATSIS, Canberra.

Altman J.C. and Taylor, L. (1987). *The Economic Viability of Aboriginal Outstations and Homelands.* AGPS, Canberra.

Berndt, R.M. (1989). *The speaking land: myth and story in Aboriginal Australia.* Penguin, Victoria.

Blanchard, C.A. (1987). *Return to Country: the Aboriginal Homelands Movement in Australia.* AGPS, Canberra.

Borrie, W.D., Smith, L.R. and Di Julio, O.B. (1975). *National Population Inquiry:Population and Australia, A Demographic Analysis and Projection. First Report, Vol. 2.* AGPS, Canberra.

Braithwaite R.W. (1991). Aboriginal fire regimes of monsoonal Australia in the 19th century. *Search,* **22**: 247-249.

Braithwaite, R.W. (1992). Black and green. *Journal of Biogeography* **19**: 113-116.

Braithwaite, R.W. (1994). *Kakadu National Park: A Case Study of Tourism in a World Heritage Area.* Tourism Conference Proceedings, Mackay, Queensland, 12-15 April 1994.

Crough, G.J., Howitt, R. and Pritchard, W. (1989). *Aboriginal Economic Development in Central Australia.* CAO, Alice Springs.

Fisk, E.K. (1985). *The Aboriginal Economy in Town and Country.* Allen and Unwin and AIATSIS, Sydney.

Foran, B. D. and Walker, B.W. (1986). *Science and technology for Aboriginal development.* CSIRO and Centre for Appropriate Technology, Alice Springs.

Horton, D. (ed) (1994). *The Encyclopaedia of Australian History.* Aboriginal Studies Press, Melbourne.

Industry Commission (1991). *Mining and Minerals Processing in Australia.* Paper No 7. AGPS, Canberra.

Lippmann, L (1981). *Generations of Resistance: The Aboriginal Struggle for Justice.* Longman and Cheshire, Sydney.

Maddocks, I. (1992). Ethics in Aboriginal research. *The Medical Journal of Australia,* **157**: 553-555

Meehan, B. (1982). *Shell Bed to Shell Midden.* AIAS, Canberra.

PATA (1992). *Endemic Tourism: A profitable industry in a sustainable environment.* Pacific Asia Travel Association, Pacific Division, Kings Cross, NSW.

Pearce, D. (1994). *Whose Rights Are They?* Ecopolitics VIII, Christchurch, New Zealand.

Resource Assessment Commission (1991). *Kakadu Conservation Zone Inquiry: Final Report.* AGPS, Canberra.

Reid, G. (1990). *A Picnic With the Natives.* Melbourne University Press, Melbourne.

Reid, J., Morton, S.R. and Baker, L. (1992). Traditional Knowledge + Ecological Survey = Better Land Management. *Search,* **23**: 249-251.

Smyth, D. (1994). *Understanding Country: The importance of land and sea in Aboriginal and Torres Strait Islander Society.* Council for Aboriginal Reconciliation, AGPS, Canberra.

Taylor, J. (1993). *The Relative Status of Indigenous Australians, 1986-91.* ANU, Canberra.

White, J.P. and Mulvaney, D. J. (1987). How many people? In: *Australians to 1788,* (eds Mulvaney, D.J. and White, J.P.), pp. 115-117. Fairfax, Syme and Weldon, Sydney.

Woodward, A.E. (1974). *Aboriginal Land Rights Commission 2nd Report.* AGPS, Canberra.

Young, E.A., Ross, H., Johnson, J., Kesteven, J. (1991). *Caring For Country — Aborigines and Land Management.* ANPWS, Canberra.

Accommodating Different Perspectives

Chapter 10

Interactions between land uses in Australia's Savannas — it's largely in the mind!

J A Taylor[1]
and R W Braithwaite[2]

[1] CSIRO Division of Tropical Crops & Pastures,
306 Carmody Road,
St. Lucia, QLD 4067
Australia

[2] CSIRO Division of Wildlife & Ecology,
Tropical Ecosystems Research Centre,
Private Bag No 44,
Winnellie, NT 0821
Australia

Abstract

This paper reviews the interactions between the major land uses in the savannas of northern Australia, and highlights the issues that influenced the structure and program of the conference. The conference organising committee's perceptions of the interactions between pastoralism, Aboriginal use, protected areas, military use, tourism, cropping and mining are considered. A matrix analysis of the nature and intensity of the interactions between land uses showed that some land uses are perceived to have both a positive and a negative impact on another land use, but overall, the number of negative interactions outweigh the positive interactions. Hence the focus of the conference on conflicts. The paper also notes that value systems underpin personal judgements about the interactions, and therefore concludes that we will not resolve most of the conflicts in managing the savanna resources. Yet, importantly, a survey of the organising committee immediately after the conference has shown that perceptions of many interactions can be influenced by multi-disciplinary meetings such as this. We should all strive to foster greater mutual understanding and respect. This will enable a more efficient arrival at satisfactory compromises by the community, and serve to maximise opportunities for collaboration.

1. INTRODUCTION

The tropical savannas are the most extensive vegetation type in northern Australia. The savannas are valued by many sectors of the community for their economic, aesthetic, recreational and spiritual benefits. For many decades, the natural resources of the region have generated considerable economic benefits to the nation ($75 billion per annum, Commonwealth of Australia 1993) through extensive land uses such as pastoralism/grazing for beef production, and intensive land uses such as mining. But over the past decade or so, community attitudes and decision making have reflected a slow but steady shift away from market values towards greater consideration of non-market values (e.g. existence and option values; Morton 1993, Holmes 1994). As a result, significant areas of the savanna have been, and are being, set aside for traditional Aboriginal uses, and for conservation and the preservation of biodiversity. However, Australia's protected area network is considered inadequate to ensure the maintenance of biodiversity and the conservation of species and communities, and sympathetic management of the intervening lands will be required to achieve these goals (Walker and Nix 1993). In recent times, tourism has also become increasingly important in the region, for a wide range of reasons (Braithwaite 1993). Tourism is seen to be both a substitute for some land uses, and a complementary land use that provides opportunities for diversification/income supplementation and educating the general public about such things as extractive industries, Aboriginal culture and conservation. Tourism is also seen by some northern Australian residents and visitors to be more environmentally benign than pastoralism and mining (Dowling 1993).

Four trends are promoting greater debate about ecological, social, cultural and ethical values, as well as traditional economic values. First, the pattern of land use is steadily changing through displacement of one land use by another (e.g. from pastoral to mining), and by attempts at multiple-use (e.g. pastoralism-tourism-conservation). Second, greater importance is being placed on maintaining savanna land use options (i.e. by retaining savanna resources in their least altered state). Third, the frequency of direct intervention, by government, has increased significantly in the last 5–10 years (Tighe and Taplin 1990). Fourth, both Commonwealth and State/Territory governments are encouraging more and more public participation in decision-making (Commonwealth of Australia 1991a, 1991b; Robertson 1994).

Each community group or sectoral interest (e.g. pastoralists, scientists, park managers, tourist operators, engineers, Aborigines) that participates brings its own biases to the debate. Each individual representing a particular community group brings perspectives that reflect their own personal experiences, their culture (i.e. the way they are taught to think about natural resources), their interactions with the land, and their social environment (Fortmann 1990, Brown and Harris 1992a, Brunson 1992, Kennedy et al. 1994).

It follows that conflict over use and management of natural resources is inevitable. However, a preliminary determination of the *level* of conflict can be made using a simple model that is based on resource renewability and whether the resources in question are a commodity or an amenity (Floyd 1993). The further the resources in question are from each other in a continuum from geo-commodity (e.g. mining), to bio-commodity (e.g. grazing for beef production) to use amenity (e.g. recreational use), to preservation amenity (e.g. conservation of biodiversity), the more intense the conflict. For example, mining in a protected area is likely to be quite contentious, whereas recreational use of grazed land is much less contentious. Similarly, reduction of livestock grazing to sustain biological diversity is likely to result in a more moderate level of conflict.

In this paper we review the issues that influenced the structure and program of the conference. In particular, we explore the interactions between various land uses, examine the value systems that underpin personal judgements about the interactions, and document changes in the organising committee's perceptions of a number of interactions, as a result of the conference. Finally, we conclude with our perceptions of the issue of managing savanna resources and resolving conflicts.

2. LAND USES

Land use in the tropical savannas is a mosaic of extensive uses (e.g. pastoralism, Aboriginal land, national parks/protected areas) in which a variety of localised and relatively intensive uses occur (e.g. tourism, mining, cropping, horticulture). The pattern and intensity of agricultural development reflects the strongly seasonal and highly variable rainfall (Taylor and Tulloch 1985, Williams et al. 1985), the low fertility and low water holding capacity of the soils (Williams et al. 1985), and the low digestibility and nutritive value of the grassy understorey (Andrew et al. 1985). The pattern of Aboriginal land and protected areas largely reflects the occurrence of vacant Crown land and land considered of low-value for any other purpose (Saddler 1980, ASTEC 1993), and in turn is reflected in the pattern of tourist visitation and developments. Mining, of course, is associated with economic deposits of minerals.

The land uses in northern Australia, ranked in order of area of savanna affected, include:

pastoralism/grazing for beef production,
Aboriginal use,
protected areas/national parks
military use,
tourism/ecotourism,
cropping/horticulture, and
mining.

However, it should be remembered that military, tourism and mining interests also have a very low intensity use over a vast area.

Forestry is an important land use elsewhere in Australia, but is insignificant in the tropical savannas because of poor soils, adverse weather conditions (especially highly variable rainfall and cyclones), and the susceptibility of the dominant eucalypts to termite attack (Haynes 1978, Lacey 1979).

The major characteristics[1] of these land uses are considered in detail in the papers dealing with specific industries or land uses. In this section we merely provide an overview and highlight trends that may be reflected in the intensity and nature of interactions between various land uses.

a) Pastoralism

A widespread (approx. 55% of the area), and year-long activity based mainly on native pastures, and the breeding and some fattening of cattle for beef production. Other livestock are insignificant. The industry comprises a population of 9500+ producers/landholders running some 7.5 million cattle or 33% of the national herd. Traditionally, pastoralism has provided employment for semi-skilled labour, but the labour input to pastoral enterprises has been declining as producers try to achieve economies in labour usage (O'Rourke et al. 1992). The gross value of production is AUD$1770 million and the gross value of exports AUD$1550 million (Source: ABS 1990-91 statistics). Overall, economic prospects for the beef industry are bright (Peterson 1994), but many pastoral enterprises in the Kimberley, Darwin and Gulf, Victoria River district, and Cape York and Gulf districts are uneconomic or only marginally viable (Wilcox and Cunningham 1994). The industry may be displaced by other users in marginal areas, and in areas of conservation and cultural significance.

b) Aboriginal use

A widespread (approx. 24% of area), year-long land use that may be quite diverse. It is based on land for which numbers of Aboriginal people still exist with a strong traditional affinity for that particular land. Its dedication to Aboriginal use has been the result of direct government proclamation or by purchase through Aboriginal organisations. Based on sites of cultural and spiritual significance and traditional hunting and collecting grounds (symbols of the deep attachment traditional Aborigines have for their land), contemporary usage may, however, include a range of market activities. The Aboriginal population of northern Australia is approximately 100 000 (ASTEC 1993) and is likely to continue increasing. The area of land is likely to continue to increase as an outcome of native title/Mabo claims. Cattle production will have a major role in future Aboriginal land use, and tourism and mining are priorities of many Aboriginal groups (Tilmouth 1994). Groups of about a hundred people engaged in a mixture of pastoral, horticultural, tourism, craft-making and traditional activities are likely to become increasingly common.

c) Protected areas/National Parks

These include reserved, dispersed tracts of land varying in size from c.1–20 000 000 ha (approx. 7% area). Although theoretically based primarily on conservation of species and communities, their dedication is often motivated by recreational and tourism needs of the community, and they are often associated with striking scenery. There is a strong overlap between conservation recognition as in World Heritage status and attractiveness to tourists. High profile is obtained through tourism promotion and through political controversy over conservation value, e.g. Kakadu National Park NT (Braithwaite and Woinarski 1990). While other habitats like wetlands, monsoon forest and rocky escarpment have been seen as most important, the savanna is increasingly seen to be important because it is rich in species (Taylor and Dunlop 1985, Braithwaite 1990) and relatively intact biologically (Woinarski and Braithwaite 1990), and it forms the matrix which is the key to the conservation management of the landscape (Braithwaite and Werner 1987). Tourism is the only acceptable commercial use of resources in protected areas (Dingwall 1992), and visitation pressure is greatest in the early dry season (June–August). The protected area is considered inadequate for conservation (Carter 1992) and is likely to increase as marginal pastoral enterprises are displaced. Many important ecosystems/biogeographic regions of the north are not currently represented in reserves (Woinarski 1992).

d) Military use

The Australian Defence Force uses a number of dispersed tracts of land (< 5% of the total area) for permanent facilities (e.g. communication, accommodation, air support infrastructure) and training exercises by the army and airforce. The intensity of use varies from regular intensive training on Defence controlled Commonwealth land (approx. 1.15 M ha) to irregular, large scale, low intensity training activities over extensive areas (e.g. the Kangaroo 92 exercise). Under the current policy of transparency, where large multi-national exercises are staged over vast areas to demonstrate Australia's military capability (Anon. 1993), military use is increasingly extensive.

e) Tourism

Tourism is a highly localised (probably < 1% area subject to intensive impact) and somewhat seasonal (dry season emphasis) land use. It is usually based on outstanding natural and cultural features, although the vastness of the savannas are an attraction in their own right. The high profile of a few places like Kakadu

---

1. Statistics are based on statistical regions which may encompass other vegetation types as well as savannas. Only those statistical regions that were dominated by savanna woodland or grassland were included.

attract international tourism while domestic tourism is more widespread. However, the impact of domestic tourism on the economy is smaller than the impact of international tourism (Harris 1992). Tourism generates more money for local people than other commercial activities, and tends to spread the economic benefits more evenly than pastoralism or mining (Dowling 1993). Its revenue is AUD$2000 million per annum; and has excellent prospects for future growth (ASTEC 1993, PATA 1992, Morton and Stafford Smith 1994). Low-key tourist developments are generally favoured by both tourists and residents alike (Dowling 1993), but predicted growth could lead to increasing demand for development in remote localities (ASTEC 1993). Tourism is the only growth area for the employment of semi-skilled labour, and is often cited as a substitute for the jobs and money that would be lost if extractive or consumptive industries ceased (Wescott and Molinski 1993). In recent years tourism in northern Australia, where the product is the natural environment, is increasingly marketed as 'green' tourism or ecotourism. This is both an activity and a philosophy (Dingwall 1992, Braithwaite 1993). It is an activity based on the use and appreciation of nature and cultural resources, and a philosophy oriented to maintaining long term sustainability of the resources on which the activity depends. Ideally, it is also a quest for personal enrichment of the tourist and improvement in the quality of the experience provided by the industry.

f) Cropping (including horticulture)

Crops include dryland grain (sorghum, maize, grain legumes, peanuts, oilseeds) and cotton, and sugar in the Ord River Irrigation Area. Overall, cropping is insignificant (<1% area), highly localised, and except perhaps for sugar in the Ord, the potential is limited. There is not likely to be any major expansion of dryland cropping. Any new areas or industries will probably be based on irrigation.

Horticultural crops include mangoes, bananas, pineapples, melons, avocados, cashews, papaya and lychee. Gross value of horticultural production is AUD$370 million per annum, and gross value of horticultural exports is AUD$160 million per annum (ASTEC 1993). There is considerable potential for growth of the industry, and exports. There is a range of tropical fruit (e.g. Alexander *et al.* 1982) increasingly penetrating the markets of the temperate world; a trend that is likely to continue. However, with increasing community disquiet about chemicals and clearing (Standing Committee on Agriculture 1991), it seems likely that this land use will face tighter controls.

g) Mining

The actual mined area is small and highly localised (<0.2% area), although mineral exploration is much more widespread. Mines such as gold mines in the Northern Territory are short-lived with an average life of about three years, so affected areas change with time. About 9% of the Northern Territory is under exploration lease, again with considerable change in the locations affected from year to year. Many mine sites in remote areas now operate on a fly-in, fly-out basis, thereby reducing their impact. Mineral exports include iron ore, gold, silver/lead/zinc, copper, bauxite/alumina, manganese, uranium, diamonds, nickel and phosphate. The gross value of production is AUD$2500 million per annum (Source: ABS 1990–91 statistics). Mining accounts for almost 90% of the merchandise exports from northern Australia (ASTEC 1993). Future access/exploration may be further restricted on conservation, environmental and Aboriginal interest grounds, and could alter dramatically with a change of government. There is likely to be significant investment in current leases.

3. USER PERCEPTIONS AND OTHER BIASES

The amount of conflict that occurs as a result of an interaction between land uses will depend largely on the scale of any differences in perceptions about a land use, and what one group thinks another group thinks about that land use (Vining and Ebreo 1991). Recent studies have highlighted major differences between residents and tourists in their perception of the impact of tourism (Dowling 1993), and between a wide range of community groups (e.g. beef producers, government advisers, bankers, teachers, environmentalists and students) in their perceptions of the impact of grazing on rangeland (MacLeod and Taylor, unpublished). For example, community groups with only an indirect association with the land, and the northern beef industry in particular, (e.g. bankers, environmentalists, teachers and students) have a significantly more negative view of the scale and intensity of grazing land degradation than those with a direct association with the land (e.g. beef producers, stock and station agents, government advisers) (MacLeod and Taylor, unpublished). The point is not that one group is misinformed or ignorant, for they are usually not (Fortmann 1990), or that one is right and the other is wrong, but that very different value systems underpin people's attitudes towards animals, natural settings, use of particular habitats (e.g. rainforest), and industry practices (e.g. clearing, prescribed fire). In some cases, community groups will have mutually exclusive value orientations (Floyd 1993). These values will evolve with time however, and hopefully lead to some convergence of the ensuing value judgements, and perhaps a compromise.

Urbanisation is progressing rapidly (Tighe and Taplin 1990). Fewer and fewer people depend directly on natural resources for their livelihood and learn about natural resource values through farming, hunting etc. More and more people learn about nature and the environment through television, books and recreational experiences such as bushwalking and bird watching. As

Table 1: The Conference Organising Committee's 1992 assessment of the positive interactions between various land uses in the tropical savannas. The nature of the perceived effect(s) are described in the text

Effect of on:	Pastoralism	Aboriginal Use	Protected Area	Military Use	Tourism	Cropping	Mining
Pastoralism		▲	●	▲	●		●
Aboriginal Use			●	▲	●		
Protected Area	●	●		▲	■	●	▲
Military Use	▲					▲	▲
Tourism	●	●	■	▲		■	
Cropping	■				●		●
Mining	▲		●		▲	▲	

where ▲ = significant positive impact
● = positive impact
■ = weak positive or almost no impact

a consequence, community values and attitudes will continue to shift away from the utilitarian view towards an environmentally oriented view of natural resources (Kennedy et al. 1994).

Many professionals have a firm conviction that their views represent the 'middle' ground, yet the few studies that have examined this issue have shown that the 'middle' ground was not where the professionals thought it was (Twight and Carroll 1983, Fortmann 1990, Vining and Ebreo 1991). For example, in Illinois the general public placed greater value on wildlife and scenery and less value on commodities (e.g. timber), than professional foresters (Vining and Ebreo 1991).

Not only do different groups have different perceptions about the nature of an issue, but they also differ in their views of what is necessary for intelligent action to resolve the conflict. Characteristically, academics (and resource managers, Brunson 1992; and many biologists, Aslin 1994) believe that the main obstacle is ignorance, and if only the problem could be explained with sufficient clarity and authority, then voters and governments would act on the scientists' advice (Young 1991). Furthermore, academics and scientists in particular, must be careful not to assume that any one discipline has a monopoly on objectivity, wisdom or value (Aslin 1994), because real conflicts deal with real-world complexity (Myers 1993, Aslin 1994), and because they themselves are inevitably strongly biased (Brunson 1992, Kennedy et al. 1994). Science must become more tolerant of other views of reality, other ways of seeing the world (Eckersley 1993). Interestingly, Brunson and Kennedy (1994) argue that land managers and scientists are really 'relationship managers', because they manage the interaction/relationship between people and the environment. They go on to point out that few people recognise this, and that in their view, this has been to the detriment of both the resources and some professions.

Even if community groups do not share the same value systems, they do share some visions of a preferred future (Brunson 1992). For example, it seems that almost everyone is concerned about, and wants to avoid, degradation of grazing land (Pitt and Yapp 1992). Brunson (1992) suggests that we can identify shared visions by broadening the realm of the debate and by learning to listen to other groups.

Interactions between land uses

In February 1992, at one of the early planning meetings, the Conference Organising Committee developed a matrix of competing land uses, and through debate, developed a consensus view of the nature and intensity of the interactions between various land uses in our tropical savannas. The outcome of this debate is shown in Tables 1 and 2

It was perceived that the interaction between land uses could result in little or no impact, or could have varying degrees of either a positive (Table 1) or a negative impact (Table 2). Clearly, these are value judgements. Perceptions of the nature and significance of the interaction will largely depend on one's value system and whether the scale of the interaction is at the level of a site (two or more contiguous or overlapping land uses, e.g. a mining operation adjacent to a national park./protected area), or a region (e.g. former pastoral land dedicated as a national park).

In our assessment, we note that although some land uses have both positive and negative effects on another land use (e.g. pastoralism and national parks, Table 1 and 2), the negative impacts, as our group scored them in early

Table 2: The Conference Organising Committee's 1992 assessment of the negative interactions between various land uses in the tropical savannas. The nature of the perceived effect(s) are described in the text

Effect of on:	Pastoralism	Aboriginal Use	Protected Area	Military Use	Tourism	Cropping	Mining
Pastoralism		■	▲	■	●	●	▲
Aboriginal Use	▲		●	■	●	●	●
Protected Area	▲	●		■	●	▲	▲
Military Use	■	●			■	■	■
Tourism	●	●	●	■		●	●
Cropping	●	●	▲	■	+		●
Mining	■		▲	■	■	■	

where ▲ = significant negative impact
● = negative impact
■ = weak negative or almost no impact

1992 (#28, Table 2), outweigh the positive impacts (#17, Table 1). No doubt other people, with different value systems, would dispute this assessment. Unfortunately, there is insufficient information and a lack of credible techniques to allow us to standardise and evaluate these interactions more fully in economic, aesthetic and cultural terms. Furthermore, while the focus of this conference (and typically the media, McIlwaine 1992, Eckersley 1993) is on the *conflicts* between land uses, it is important that we do not lose sight of the positive interactions between the various land uses in the savanna region. Obviously it is in everyone's interest to maximise the positive interactions whenever possible.

Positive interactions between land uses

Taking an overview of Table 1, we note that it was perceived that national parks, tourism and pastoralism all provide significant positive benefits to other sectors, and that mining, Aboriginal use and military use provide relatively small benefits to other users.

Pastoralism

Pastoralism has provided infrastructure (albeit basic in many areas) for tourism, military use and mining; employment and income for some members of Aboriginal communities; local markets for feed grain and forages from cropping land; landscape and scenic diversity which may enhance tourist appeal; and habitat modifications that may enhance faunal diversity and the population and visibility of some fauna (e.g. macropods through provision of waters, and heavily grazed grassland). Other infrastructure has been provided by government to assist the pastoral industry (e.g. beef roads) and is now used by other sectors.

Aboriginal use

Aboriginal people were heavily employed in the cattle industry and still have a substantial presence. More recently, Aboriginal people have found applications for their expertise in ecology and ethnobotany in land/habitat management in national parks and other protected areas, and in tourism ventures. Some find employment in various aspects of mining. But to be realistic, employment opportunities for Aborigines in the private sector have been very limited and the bulk of Aboriginal employment is financed by the government sector (Altman 1980). More recently, the culture and craft of many communities has become a significant tourist attraction in its own right. Aboriginal culture is seen as one the main drawcards of Australian tourism, particularly in the international marketplace (PATA 1992).

Protected area

The presence of protected areas or national parks tends to enhance the conservation ethic in a region and usually leads to some improvement in infrastructure (e.g. roads and camping facilities) that benefits both neighbours (e.g. pastoralists) and tourists. In Queensland, national parks in the savanna region also provide for some livestock grazing. National parks provide employment for Aboriginal people in particular as rangers and cultural advisers (40% of staff of Kakadu National Park are Aboriginal, P. Wellings, *pers. comm.*), and these and other park staff may be a source of information on native vegetation for rehabilitation of mined areas, and habitat management for pastoralists interested in management for production and conservation.

Military Use

The positive impacts of military use arise from a consultative approach to activities on non-defence land, including compliance with conditions of access and compensation for damages incurred during exercises. There may also be significant contributions to local economies through construction projects and purchase of provisions during exercises.

Importantly, Defence land could afford protection for some endangered species because of the restricted access, compliance with environmental management guidelines and the nature of many Defence activities.

Tourism

Increasing numbers of tourists are associated with increasing public interest in natural areas, conservation and Aboriginal culture. Visitation further enhances public support for conservation and national parks. Tourism provides employment for some Aboriginal people, thus enhancing their pride in their culture and their feeling of self worth, but it also provides a growing source of revenue through the sale of artefacts. This in turn has ensured the survival of traditional Aboriginal skills and crafts in many areas. Tourism also provides an opportunity for mining and pastoral enterprises to educate the public and enhance their public image. For some pastoralists, tourism can also constitute a significant income supplement as more and more tourists seek to experience the many facets of outback life and Australian lifestyle (another of the major assets of Australian tourism, PATA 1992). Cropping and horticultural enterprises may benefit from increased local sales of produce, and perhaps through supplying a growing but limited demand for local produce from accommodation centres and restaurants.

Cropping

The cropping and horticultural industries provide some foodstuffs for the passing tourist trade and accommodation centres, but also food for livestock (fodder and grains), and inadvertently for some fauna (e.g. parrots, fruit bats). The cropping enterprises themselves may sometimes be a tourist attraction, and to some may enhance the diversity of the landscape and its visual appeal. Cultural practices and techniques may find application in rehabilitation of mined and degraded sites.

Mining

Mining has led to very significant improvements in infrastructure (e.g. roads, railways, transport, health and communications) that are of benefit to pastoralists, Aboriginal communities, national parks, tourism and the military. Indeed, mining royalties are a significant source of revenue for some Aboriginal communities. For example, they have enabled the purchase of many of the tourism ventures based in Kakadu. Mining also provides some opportunities for employment and skill development by Aboriginal and other local people, and may provide a market for some local produce. Mining sites are of interest to some of the general public, and the industry help provide a diverse tourist/educational experience for visitors to the region.

Another example of mutual benefit across industries is in research. For example, restoration research is of value to pastoralism, mining, tourism, the military and others. A review of cross-sectoral research needs for the savanna would be timely. More progressive funding bodies (e.g. Land and Water Resources R&D Corporation) are increasingly encouraging and funding multi-sectoral research. The ESD Inter-sectoral Issues Report makes a range of recommendations for the research area (Commonwealth of Australia 1992).

4. CONFLICTS BETWEEN LAND USES

Taking an overview of Table 2, it is interesting to note that national parks were perceived to have varying levels of negative impact on all other land uses in the region. The land uses that were identified by the Committee as having a significant negative impact on two or more other land uses were pastoralism, national parks and mining. As in Table 1, the negative impact of the military was seen to be weak overall.

Pastoralism

Pastoral lands are sometimes seen by other land managers as a harbour for weeds and feral animals, and some pastoral practices (e.g. pasture improvement) are seen to be a major source of environmental weeds (Humphries *et al.* 1991, Lonsdale 1994). Furthermore, some of the other practices used on pastoral land (e.g. baiting to control predators, herbicides, burning) may, if used indiscriminately, have a serious effect on the diversity and abundance of fauna and flora. Without sympathetic management, grazing and other pastoral practices could also damage sites of scenic and cultural value, and adversely affect habitat quality for a range of ground-dwelling animals, particularly in the habitats adjacent to water. Furthermore, poor drought management could seriously damage the image of the pastoral industry, especially in the eyes of Australian and international tourists.

Aboriginal use

Aboriginal land is seen by many to effectively reduce the area of 'productive' land (i.e. crop and pastoral land) and to impose severe restrictions on tourist access, and mineral exploration and mining. Concern has been expressed that Aboriginal lands harbour weeds, feral animals and livestock diseases, especially if livestock and feral animal management on Aboriginal lands is poor. Concern has also been expressed that Aboriginal land could also be a source of wildfire, which would put Aboriginal land management practices into conflict with pastoralists and some park rangers.

Table 3: The Conference Organising Committee's 1994 assessment of the positive interactions between various land uses in the tropical savannas. The nature of the perceived effect(s) are described in the text

Effect of on:	Pastoralism	Aboriginal Use	Protected Area	Military Use	Tourism	Cropping	Mining
Pastoralism		●	■	■	●	●	●
Aboriginal Use	■				●		●
Protected Area	■	●		●	▲		
Military Use							
Tourism	●	●	▲	■		■	●
Cropping	●				●		■
Mining	■		■		■	■	

where ▲ = significant positive impact
● = positive impact
■ = weak positive or almost no impact

Protected areas

The major sources of conflict over protected areas are the views that they restrict the use of otherwise 'productive' land, especially by livestock and mining. National parks are also seen by many agriculturalists as a harbour for pests (weeds and fauna). Visitors to national parks are often frustrated by restrictions on where they can go. Increasingly national parks have collaborative management arrangements with indigenous people (West and Brechin 1991). This is increasingly happening in northern Australian parks under a variety of models (Birckhead and Smith 1992). However, there are areas of conflict between traditional national park conservation values and Aboriginal aspirations and these require compromises by both sides (Braithwaite 1992). For example, in Kakadu the traditional Aboriginal owners' wish to hunt and to maintain herds of feral buffalo in the park, but the conservationists' values are compromised by this. There is one tightly controlled and well managed herd in a fenced area of the Park.

Military use

Areas dedicated solely to military use have very stringent restrictions on access. The negative effects of military use on adjoining lands is largely through damage to infrastructure (e.g. roads) and the off-site effects of vehicles and live firing on the occurrence and intensity of erosion and fire, and perhaps disturbance of livestock and fauna. Military activities have the potential to introduce and spread noxious and environmental weeds and exotic fauna, such as cane toads, on vehicles, personnel and equipment.

Tourism

Tourists are often regarded by local inhabitants as a nuisance because they cause traffic and overcrowding problems, and are a major source of litter and vandalism (Dowling 1993). Tourists have affected habitat value and desecrated sites of cultural and environmental significance directly through trampling, disturbance of fauna and flora, littering, and noise and water pollution, and indirectly by spreading weeds and pests. Indiscriminate use of 4WD vehicles can have a devastating impact (Buckley and Pannell 1990), particularly if conditions make trafficability difficult. Tourists also bring with them the problems of consumerism — alcohol, petrol sniffing — and serious diseases which have now become prevalent in Aboriginal communities. However, tourists' perceptions, generated by their experiences in the savanna, could have a significant long-term impact on public views, and ultimately on political action in connection with conservation, mining, cropping and pastoralism.

Cropping

Conflict relating to cropping and horticulture largely relates to practices involving cultivation and chemicals, and subsequent siltation and pollution of waterbodies. In the seasonal dryness of the wet-dry tropical climate of the savanna lands, any industry which uses groundwater is potentially impacting on the others, including conservation (Braithwaite 1994). Irrigation is a major user of this resource. Crop lands are also seen to be a source of weeds, and to the detriment of fauna, hold an often fatal attraction. This may encourage abnormal variations in the abundance and seasonal distribution of animals such as granivores and frugivores (Andrew et al. 1985).

Table 4: The Conference Organising Committee's 1994 assessment of the negative interactions between various land uses in the tropical savannas. The nature of the perceived effect(s) are described in the text

Effect of on:	Pastoralism	Aboriginal Use	Protected Area	Military Use	Tourism	Cropping	Mining
Pastoralism		•	•	▲	▲	▲	•
Aboriginal Use	■		▲	▲	▲	•	•
Protected Area	■	▲		•	▲	■	■
Military Use	▲	▲	•		▲	•	▲
Tourism	•	•				▲	▲
Cropping	▲	•	•	▲	▲		•
Mining	▲	■	■	▲	▲	▲	

where ■ = weak negative or almost no impact
● = negative impact
▲ = significant negative impact

Mining

Mining can detract from the scenic value of a landscape, and unsympathetic access to areas could directly and indirectly damage their cultural and environmentally sensitive values. Inappropriate siting of tracks, and use of 4WD vehicles and trail bikes could lead to increased erosion, and together with the influx of people and their pets, could facilitate the introduction and spread of a range of pests. However, the chances of this happening today are much reduced through legislation, the requirement to prepare environmental impact assessments, and a changing attitude in the industry. Nevertheless, mining is considered to be incompatible with national park and protected area values. Other perceived negative impacts of mining include vandalism and littering. The type of mineral to be extracted (e.g. uranium cf. silver/lead/zinc) can have a bearing on the intensity of opposition to mining (Saddler 1980).

5. Changes in Perception Over Time

The preceding sections on positive interactions and conflicts between land uses were based on the Conference Organising Committee's perceptions as at March 1992. By the time of the conference (July 1994), several members of the Committee felt that some of their perceptions had changed. Other members were influenced by the pre-conference tour (which exposed them to aspects of military use, mining, protected areas and pastoralism in the savannas), as well as the presentations and workshops that were part of the conference. To document these changes in perception over the two and a half years, those who were on the Organising Committee in early 1992, and who attended the conference, were asked to redo the interaction matrix. A collation of the responses (8) is shown in Table 3 and 4.

In 1992 the Committee identified 17 positive interactions and 28 land use conflicts in the savannas (Table 1 and 2). In August 1994 they identified almost the same number of positive interactions (16, Table 3), but their perception of the number of conflicting situations had fallen dramatically from 28 to 18 (Table 4). The reasons for these changes are complex and we are unable to address them here. What is important to note is that the Committee's perceptions have evolved significantly with time, experience, and perhaps the presentations and interactions at the conference. Three 'new' interactions were identified in 1994 — a positive impact of cropping on pastoralism, and positive effects of mining on Aboriginal use and tourism (Table 3). Of the positive interactions identified in 1992, the intensity of six were downgraded (the effect of pastoralism on protected areas, military use and cropping, the effect of protected areas on pastoralism and mining, and the effect of cropping on protected areas — Table 3), and five were upgraded in 1994 (the effect of Aboriginal use on pastoralism, the effect of military use on protected areas, the effect of cropping on pastoralism, and the effect of mining on aboriginal use and tourism - Table 3). Of the negative interactions or conflicts identified in 1992, the intensity of 14 were downgraded (effect of pastoralism on cropping, effect of aboriginal use on protected areas, effect of tourism on pastoralism, aboriginal use, protected areas and cropping, the effect of cropping on pastoralism, military use and tourism, the effect of mining on pastoralism and tourism, and the effect of protected areas on aboriginal use, tourism and cropping — Table 4), but *none* were upgraded in 1994 (Table 4).

These changes reinforce the points that perceptions evolve, but perhaps more importantly, that people's perceptions of the intensity of conflict often diminish with experience, and of course, with time.

6. CONCLUSIONS

A major change in attitude towards land development has swept the world, with society shifting from a commodity and economic development orientation to an environmentally-oriented view of natural resources (Quigley 1989, Brown and Harris 1992b, Morton 1993, Holmes 1994, Kennedy et al. 1994). Nevertheless, there will always be competitive demands for resources. Some of these demands will be expressed through the market system, some will require the creation of a political process (e.g. the Resource Assessment Commission) to act as a proxy market. With increasing urbanisation and ethnic shifts, it is clear that Australian society (and its needs and values) will be very different in the year 2000 to what it was in the 1960s.

For land users and managers to respond pro-actively to government initiatives or the criticism of interest groups, they need to understand how the opinions of the general public match those of the more vocal interest groups (Brunson and Steel 1994), and what the general public thinks about the state of the savanna and savanna management issues. We have provided some insights, but this is an area that warrants considerable attention. Aslin (1994), in writing about conservation of biodiversity, has argued that research on people's values and attitudes is even more important and more urgent than ecological and taxonomic research. This is because people's values, and the choices that they lead to, usually determines how species and their habitats (and the land in general) are used, rather than biology or economic information *per se* (Aslin 1994).

Considerable progress has been made toward better land management through community action groups like Landcare and Integrated Catchment Management. With both Commonwealth and State governments now committed to further community participation in land use planning and management through national strategies such as the National Strategy on Ecologically Sustainable Development (Commonwealth of Australia 1991b), the National Strategy for the Conservation of Biodiversity (ANZECC Task Force on Biological Diversity 1994), and the National Rangeland Strategy (Robertson 1994), community pressure for responsible land use and management is likely to increase significantly. Conflict seems unavoidable. But it is not something to be afraid of: it is a symptom of change and growth in society (Floyd 1993), and perhaps even a healthy sign of a democracy in action (McDonald 1992). We suggest that the major conflicts will continue to be between consumptive or market uses by individuals or private industry (e.g. pastoral and mining) and community or government needs to conserve biodiversity and Aboriginal culture, and particularly where land is managed for multiple use. To be more specific, we believe it will be largely a conflict between agriculture and urban values about the use of nature; it will not be a conflict of facts.

Ultimately, multiple land use is the future of the Australian savannas. Its success will depend on making the most of the positive interactions between industries, and dealing efficiently and accepting the need for compromises with the negatives. To aid the process of reconciliation of different interest groups and the efficient arrival at appropriate compromises, governments may need to put appropriate structures/mechanisms in place (Commonwealth of Australia 1992). The Resource Assessment Commission (RAC) was an important attempt to provide what was needed to deal with such issues. While there are always going to be critics or sectional interests who cry foul, the RAC was widely seen as impartial, thorough, multi-disciplinary and knowledgeable (Irving Saulwick and Associates 1993). Presided over by a judge, and with an economist and an ecologist as co-commissioners plus good technical support, the RAC was able to provide the public and the government with a clear analysis of complex issues, allowing efficient and transparent decision-making by government to occur.

We concur with Young (1991) that the main obstacle to intelligent action to resolve land use conflicts is not ignorance or lack of understanding, but the conflict of interest and values. Hardin (1968) would probably characterise it as an example of the 'tragedy of the commons', where the short-term gain by a sector from overuse of our common heritage is balanced against the somewhat smaller per capita loss by the whole community. It follows that while decisions are made, we will probably not totally resolve most of the resource management conflicts in the savanna, or in any other region for that matter. The challenge is to be more constructive, to negotiate ways of moving forward that all parties will accept. However, the reality is that the group (or individual) with the most power will dominate the decision-making process, and the underlying value-based problem will remain and continue to rankle some. The real challenges for resource managers then, will be a) to understand more about the evolving values and value systems of various community groups, b) to help particular groups assert their values, c) to negotiate ways of moving forward in the face of competing demands for resources, and d) to do their best in managing the conflict and animosity that results.

Acknowledgements

We thank Andrew Ash, Alan Barton, Peter Cox, Andrew Johnson, Mark Lonsdale, Sue McIntyre and John McIvor for valuable comments on a draft of the paper.

References

Alexander, D. McE., Scholefield, P.B. and Frodsham, A. (1982). *Some Tree Fruits for Tropical Australia*. CSIRO, Australia.

Altman, J.C. (1980). The Aboriginal economy. In: *Northern Australia: Options and Implications*, (ed Jones, R.), pp. 87-107. ANU Press, Canberra.

Andrew, M.H., Gowland, P.N., Holt, J.A., Mott, J.J. and Strickland, G.R. (1985). Vegetation and fauna. In: *Agro-research for the semi-arid tropics : North-west Australia*, (ed Muchow, R.C.), pp. 93-111. University of Queensland Press, Brisbane.

Anon (1993). *Strategic Review 1993*. Dpubs, Defence Centre, Canberra.

ANZECC Task Force on Biological Diversity (1994). *Draft National Strategy for the Conservation of Australia's Biological Diversity*. AGPS, Canberra.

Aslin, H.J. (1994). Values and attitudes to biodiversity - diverse views. *Australian Biologist*, **7**: 49-57.

Australian Science and Technology Council (ASTEC) (1993). *Research and Technology in Tropical Australia and their application to the development of the region*. AGPS, Canberra.

Birckhead, J.T. de Lacy and Smith, L. (1992). *Aboriginal Involvement in Parks and Protected Areas*. Aboriginal Studies Press, Canberra.

Braithwaite, R.W. (1990). Australia's unique biota: implications for ecological processes. *Journal of Biogeography*, **17**: 347-54.

Braithwaite, R.W. (1992). Black and Green. *Journal of Biogeography*, **19**: 113-6.

Braithwaite, R.W. (1993). Ecotourism in the monsoonal tropics. *Issues*, **23**: 29-35.

Braithwaite, R.W. (1994). *Working with a recalcitrant land: Maintaining conservation value and improving economic production of Australia's northern lands*. Australian Rangelands Society 8th Biennial Conference Proceedings/Working Papers, pp. 89-96.

Braithwaite, R.W. and Werner, P.A. (1987). The biological value of Kakadu National Park. *Search*, **18**: 296-301.

Braithwaite, R.W. and Woinarski, J.C.Z. (1990). Coronation Hill, Kakadu Stage III — assessing the conservation value. *Australian Biologist*, **3**: 3-13.

Brown, G.G. and Harris, C.C. (1992a). The USDA Forest Service: toward the new resource management paradigm? *Society and Natural Resources*, **5**: 231-245.

Brown, G.G. and Harris, C.C. (1992b). The Forest Service: Changing of the guard. *Natural Resources Journal*, **32**: 449-466.

Brunson, M.W. (1992). Professional bias, public perspectives and communication pitfalls for natural resource managers. *Rangelands*, **14**: 292-295.

Brunson, M.W. and Kennedy, J.J. (1994). Redefining 'multiple use': agency responses to changing social values. In *A New Century for Natural Resources*. (eds Knight, R.L. and Bates, S.F.), (in press).

Brunson, M.W and Steel, B.S. (1994). National public attitudes towards federal land management. *Rangelands* (in press).

Buckley, R. and Pannell, J. (1990). Environmental impacts of tourism and recreation in national parks and conservation reserves. *Journal of Tourism Studies*, **1**: 24-32.

Carter, L.E (1992). Wilderness and its role in the preservation of biodiversity: the need for a shift in emphasis. *Australian Zoologist*, **28**: 28-36.

Commonwealth of Australia (1991a). *Decade of Landcare plan - Commonwealth component*. AGPS, Canberra.

Commonwealth of Australia (1991b). *National Strategy for Ecologically Sustainable Development*. AGPS, Canberra.

Commonwealth of Australia (1992). *Ecologically Sustainable Development Working Group. Intersectoral Issues Report*. AGPS, Canberra.

Commonwealth of Australia (1993). *Research and Technology in Tropical Australia: Selected Issues*. AGPS, Canberra.

Dingwall, P.R. (1992). Tourism in protected areas: conflict or saviour. *Proceedings of the Royal Australian Institute of Parks and Recreation*, **28**: 117-122.

Dowling, R.K. (1993). Tourist and resident perceptions of the environment-tourism relationship in the Gascoyne Region, Western Australia. *GeoJournal*, **29**: 243-251.

Eckersley, R. (1993). The West's deepening cultural crisis. *The Futurist*, **27**: 8-12.

Floyd, D.W. (1993). Managing rangeland resources conflicts. *Rangelands*, **15**: 27-30.

Fortmann, L. (1990). The role of professional norms and beliefs in the agency-client relations of natural resource bureaucracies. *Natural Resources Journal*, **30**: 361-380.

Hardin, G. (1968). The tragedy of the commons. *Science* **162**: 1243-8.

Harris, P. (1992). The economy of northern Australia. In: *Research and Technology in Tropical Australia - Symposia*. ASTEC Occasional Paper No. 23. AGPS, Canberra.

Haynes, C.D. (1978). Land trees and man (Gunret, gundulk, djabining). *Commonwealth Forestry Review* **57**: 99-106.

Holmes, J. (1994). Changing rangeland resource values - implications for land tenure and rural settlement. In: *Outlook 94*, pp. 160-175. ABARE, Canberra.

Humphries, S.E, R.H. Groves & D.S. Mitchell. (1991). Plant invasions: the incidence of environmental weeds in Australia. *Kowari*, **2**: 1-127.

Irving Saulwick and Associates (1993). *Survey of Opinions about the Work of the Resource Assessment Commission*. Occasional Publication No. 3, RAC. AGPS, Canberra.

Kennedy, J.J, Fox, B.L. and Osen, T.D. (1994). Changing social values and images of good public rangeland management in the last half century. *Rangelands* (in press).

Lacey, C.J. (1979). Forestry in the Top End of the Northern Territory: Part of the Northern Myth. *Search*, **10**: 174-180.

Lonsdale, W.M. (1994). Inviting trouble: introduced pasture species in northern Australia. *Australian Journal of Ecology*, **19**: 345-354.

McDonald, G. (1992). Minimising conflict by land-use planning. In: *Resolving Disputes in the Public Interest*, (ed Newmann, R.), pp. 107-126. Environment Institute of Australia, Brisbane.

McIlwaine, S. (1992). Media. In *Resolving Disputes in the Public Interest*, (ed Newmann, R.), pp. 85-94. Environment Institute of Australia, Brisbane.

Morton, S.R. (1993). Changing conservation perceptions in the Australian rangelands. *The Rangeland Journal*, **15**: 145-153.

Morton, S.R. and Stafford Smith, D.M. 1994). Alternative land uses in the rangelands. In: *R & D for Sustainable Use and Management of Australia's Rangelands*, (eds Morton, S.R. and Price, P.), pp. 29-30. LWRRDC Occasional Paper Series No 06/93, LWRRDC, Canberra.

Myers, N. (1993). The question of linkages in environment and development. *BioScience* **43**: 302-309.

O'Rourke, P.L., Winks, L. and Kelly, A.M. (1992). *North Australia beef producer survey 1990*. Queensland Department of Primary Industries and Meat Research Corporation, Brisbane.

PATA (1992). *Endemic Tourism. A Profitable Industry in a Sustainable Environment* Pacific Asia Travel Association, Pacific Division, Sydney.

Peterson, D. (1994). Farm financial performance: outlook and analysis. In: *Outlook 94*, pp. 176-184. ABARE, Canberra.

Pitt, M. W and Yapp, T.P. (1992). Perceptions of land degradation and awareness of conservation programs in north-eastern NSW. *Proceedings 7th ISCO Conference*, 1: 115-124.

Quigley, T. M. (1989). Value shifts in multiple use products from rangelands. *Rangelands*, 11: 275-279.

Robertson, G. (1994). A national rangeland management strategy. In: *Outlook 94*, pp. 176-184. ABARE, Canberra.

Saddler, H. (1980). Implications of the battle for the Alligator rivers; Land use planning and environmental protection. In *Northern Australia: Options and Implications*. (ed Jones, Rhys), pp. 187-200. ANU Press, Canberra.

Standing Committee on Agriculture (1991). Sustainable Agriculture. SCA Technical Report Series No 36.

Taylor, J.A and Dunlop, C.R. (1985). Plant communities of the wet-dry tropics of Australia: the Alligator Rivers region. *Proceedings of the Ecological Society of Australia*, 13: 83-127.

Taylor, J.A and Tulloch, D. (1985). Rainfall in the wet-dry tropics: Extreme events at Darwin and similarities between years during the period 1870-1983 incl. *Australian Journal of Ecology*, 10: 281-295.

Tighe, P. and Taplin, R. (1990). Australian environmental politics: An analysis of past and present issues. In *Changing Directions*, (eds Dyer, K. and Young, J.), pp. 102-116. University of Adelaide Printing, Adelaide.

Tilmouth, T. (1994). Aboriginal aspirations for land management. In: *R & D for Sustainable Use and Management of Australia's Rangelands*, (eds Morton, S.R. and Price, P.), pp. 22-24. LWRRDC Occasional Paper Series No 06/93. LWRRDC, Canberra.

Twight, B.W. and Carroll, M.S. (1983). Workshops in public involvement: Do they help find common ground? *Journal of Forestry*, 81: 732-735.

Vining, J. and Ebreo, A. (1991). Are you thinking what I think you are? A study of actual and estimated goal priorities and decision preferences of resource managers, environmentalists and the public. *Society and Natural Resources*, 4: 177-196.

Walker, B. H & Nix, H. A (1993). Managing Australia's biological diversity. *Search*, 24: 173-178.

Wescott, G. and Molinski, J. (1993). Loving our parks to death: a cautionary tale. *Habitat Australia*, 21: 14-19.

West, P.C. and Brechin, S.R. (1991). *Resident Peoples and National Parks*. University of Arizona Press, Tucson.

Williams, J., Day, K.J., Isbell, R.F. and Reddy, S.J. (1985). Soils and climate. In *Agro-research for the semi-arid tropics: North-West Australia*, (ed Muchow, R.C.), pp. 31-92. University of Queensland Press, Brisbane.

Wilcox, D.G. and Cunningham, G.M. (1994). Economic and ecological sustainability of current land use in Australia's rangelands. In *R & D for Sustainable Use and Management of Australia's Rangelands*, (eds Morton, S.R. and Price, P.), pp. 87-171. LWRRDC Occasional Paper Series No 06/93. LWRRDC, Canberra.

Woinarski, J.C.Z. (1992). Biogeography and conservation of reptiles, mammals and birds across north-western Australia: an inventory and base for planning an ecological reserve system. *Wildlife Research*, 19: 665-705.

Woinarski, J.C.Z. and Braithwaite, R.W. (1990). Conservation foci for Australian birds and mammals. *Search*, 21: 65-68.

Young, J. (1991). *Sustaining the Earth*. NSW University Press, Randwick.

Chapter 11

Mediation and Sustainable Development: Conflict Prevention, Conflict Resolution and Public Participation

Donna Craig

Craig & Ehrlich
Solicitors
Level 9, 1 Chifley Square
Sydney NSW 2000
Australia

Abstract

The integration of economic, ecological social and cultural dimensions of sustainable development requires processes — it will not happen automatically. Mediation is discussed as an example of an integrating approach. The major emphasis, in the paper, is on the mediation of policy processes in a way which optimises public participation. A distinction is made between mediating specific disputes, involving limited parties and issues and mediating the processes for developing policy options. The former approach to mediation is primarily concerned about dispute resolution. The latter approach to mediation is directed towards the following objectives:

- *preventing conflict where possible;*
- *identifying (scoping) important issues;*
- *developing participatory processes to understand the issues and to develop relevant policy options.*

These procedures can be mediated as well as outcomes. The preventative and proactive form of mediation is most appropriate in integrating social and cultural dimensions of sustainable development policy relevant to tropical savannas.

1. INTRODUCTION

This paper adopts a wide definition of sustainable development and related mediation approaches. It discusses the essential social and cultural dimensions of sustainable development. Existing policy processes are poorly suited to understanding and valuing these types of factors. In the policy sciences they are often characterised as 'wicked' or intractable problems. In political practice, they are conveniently ignored in favour of quantifiable economic and scientific factors.

However, a tenacious public has never let these concerns drop from the agenda. Social and cultural issues have been asserted through a wide variety of processes such as public participation, environmental impact assessment and social impact assessment. They have achieved a new arena through the concept of sustainable development.

2. SUSTAINABLE DEVELOPMENT — DEFINITIONS, PRINCIPLES AND PROCESSES

'Sustainable development', the environmentally sound catch phrase of seemingly universal appeal, was first popularised by the Brundtland Report (World Commission on Environment and Development 1990). Ultimately, sustainable development is the integration of environment and economy into decision-making at the highest level of government policy. However, the *'beguiling simplicity and apparently self-evident meaning have obscured (sustainable development's) inherent ambiguity'* (O'Riordan 1988) such that, taken out of context of the Brundtland Report, *'sustainable development' condones an interpretation that responsibility for our environment is achievable without sacrificing economic growth*'[1] It is therefore important to analyse sustainable development in terms of broad based principles inherent to its realisation and strategies for its operationalisation. The viability and value of these principles and practices can then be critically reviewed to see if new forms of sustainable and ecologically directed change are emerging.

Although 'sustainability' was first used within the context of the sustainable management or utilisation of living resources, the Brundtland Report expanded the reference to the socio-economic realm "*where the goal is not a sustained level of physical stock or physical production from an ecosystem over time, but some sustained increase in the level of societal and individual welfare*" (Dixon and Fallon 1988). This broader context was generated by the Brundtland Report because of its concern to relieve poverty accelerating in developing countries.

In Canada, the British Columbia Task Force on Environment and Economy has suggested that the following principles are fundamental to sustainable development:

- **Economic sustainability**
  development should stabilise, and where necessary expand, the resident labour force, and help to expand and stabilise the economy.

- **Social sustainability**
  Development should help to maintain, and if possible strengthen, community identity by increasing compatibilities and reducing conflicts among the different interest groups.

- **Cultural sustainability**
  Development should be compatible with, and supportive of, the local culture and values.

- **Biological sustainability**
  Development should be compatible with the maintenance of ecological resources, biological diversity, and the harvesting of resources at sustainable levels.

- **Ethics**
  The needs and values of all interest groups should be respected and considered. In addition, people have an ethical responsibility for other species.

- **Future options**
  Policy decisions that do not foreclose future options should be given precedence; they should be responsive to the need for change.

- **Cross-sectoral approach**
  The effect of proposed development on other sectors and interest groups should be given full account. Resource management should identify common interests and compatibilities among sectors, so that the integration of the variety of interests is ensured wherever possible.

- **Optimal uses**
  Some resource uses are incompatible with others, and consequently a choice must be made. In determining the best use of land, its ability to sustain uses and resources should guide decision-makers in the context of environmental, economic, cultural and social values.

- **Consensus and common vision**
  Local governments and communities should play a greater part in setting the agendas for economic development and environmental conservation. The public should be involved in decisions concerning natural resources and environment, from the definition of objectives to a full review of policies, programmes and projects.

---

1. It is for this reason that Boer asserts that it is more appropriate to speak of 'sustainability' than 'sustainable development' 'in order to avoid the exploitative connotation in the unexamined use of the world 'development'" (Boer 1992). In this paper I will generally use the term 'sustainable development', being most widely recognised. It is also important to note the Australian Government's interpretation of sustainable development is 'ecologically sustainable development'.

The use of these principles is a way of recognising and defining the complex issues involved in order to understand sustainable development in tangible, realisable terms which can begin to overcome the ambiguity of the generic 'sustainable development'.

The principles represent a diversity of economic, ecological, social and cultural interests and values-all of which are essential to sustainable development. For the realisation of sustainable development all principles fundamental to it must be equally recognised and integrated to create a 'culture of sustainability' (Boer 1992). Cameron (1989) also recognised the features inherent in sustainable development: *'Distinction needs to be made between* physical sustainability, *which is the ability to physically support a human population in a lasting manner,* ecological sustainability, *which focuses on organising human activity to minimise ecological disruption,* and social sustainability, *which aims at such goals as social equality and maintenance of traditional societies'.*

Sustainable development cannot be achieved without the integration of physical, ecological, economic and social strategies. This integration requires a sophisticated use of mediation and public participation (in its many forms). This paper considers some of the strategies and experiences which will be useful in developing sustainable development strategies for the tropical savannas.

3. FORMS OF MEDIATION

Mediation has been described as *'a voluntary process in which those involved in a dispute jointly explore and reconcile their differences. The mediator has no power to impose a settlement. His or her strength lies in the ability to assist the parties in resolving their own differences'* (Susskind and Weinstein 1980).

Mediation may be assisted (by a third party) or unassisted. The latter form of dispute resolution is often described as negotiation. I prefer to define mediation as assisted, or unassisted, forms of dispute prevention and resolution. In essence, mediation is a process whereby conflicting parties come together to identify the issues causing conflict (move from 'position' to 'interest'[2]) and also attempt to generate options that incorporate the important interests of all of the stakeholders.

There is a perception of successful mediation as a process which necessitates compromise. Such an argument, however, denies the value of mediation in defining the interests or issues underpinning the dispute, an approach which may facilitate a new and valuable perspective for the parties to the dispute. Thus, the true value of mediation is commonly argued to lie in its ability to address the substantive issues of a dispute, thus potentially expanding the parameters delineating potential agreement (Amy 1990).

Another use of mediation is to narrow the focus of negotiation so that the central and important issues are discussed by the parties. Most importantly, mediation of sustainable development strategies can go beyond 'dispute' settlement and compromise as commonly understood. It can enhance the resources and contributing perspectives brought to bear in policy-making processes, policy options and their implementation.

There may be times when compromise is appropriate. However, compromise results in the parties walking away with a series of 'wins' and 'losses' and, to an extent, it prevents the formulation of new solutions better incorporating the needs of the stakeholders: *'If compromise is going to be acceptable to the parties it is only going to be so to the extent to which it does not abrogate values which the parties hold as non-negotiable'* (Boer et al. 1991a)[3].

Thus, even if the compromise occurs, an impasse may block the process if stakeholders reach positions which they perceive to be non-negotiable. It is, at this point that the mediation process may be applicable.

If mediation does not produce an outcome/decision acceptable to all (or any) of the parties, the process itself can still be valuable. The process of mediation has the advantage of identifying the principle stakeholders. Even if it does not identify *all* the stakeholders, it facilitates greater inclusion of interests than standard decision-making processes (or adjudicative techniques) and the identification and reduction of the number of issues in conflict so that maximum resources and effort can be devoted to the most important ones. It also has an educative function: *'it educates the parties as to the other disputants' positions, thus improving the relationship between the parties and perhaps reducing the likelihood or magnitude of future disputes'* (Kubasek and Silverman 1988). Most significantly, mediation can provide a means of integrating expert knowledge, values and policy needs.

Thus, even an 'unsuccessful' mediation potentially extends greater understanding of the conflict to the ultimate decision-maker who may then be in a position to make a 'better' decision. In addition, if the decision-maker respects the information generated by the mediation, the parties have had some degree of participatory

---

2. Mediation facilitates the movement from 'position' to 'interest'. That is it removes the emphasis from the dispute (or potential point of conflict) which arises when parties have overtly conflicting stands on an issue to the reasons behind the stand they are adopting. Once interests are identified, the parameters defining an acceptable solution are likely to have expanded or, at least, clarified.

3. It is important to note that proponents of social and cultural issues are more likely to hold non-negotiable, bottom line positions than those of economic and physical (environmental) issues, which cannot be subject of compromise and may limit the scope of solutions available through mediation.

input into the decision-making process, which should allow a more enduring decision to be reached.

An important value of mediation in the context of sustainable development is its facilitation of a mechanism for greater citizen participation at the many levels policy formulation and implementation. By encouraging a decision-making process which respects diversity and active citizenship there can be creative and practical strategies developed which constitute real change and not more of the same. In the context of sustainable development policy, it must have democratic direction. Mediation does not substitute for a democratic process, in the wider sense, but mediation processes can enhance participatory experiences and develop further policy options.

Although social and cultural values are widely accepted as relevant to the realisation of sustainable development (see the principles outlined above) they are not yet adequately recognised or integrated into the decision-making process (Craig 1989). An enduring example of failure to appreciate the social and cultural dimensions of sustainability (equally with economic and ecological dimensions) is the treatment of indigenous society and cultural values.

The social infrastructure, cultural values and the relationship to land of indigenous peoples do not easily align with the dominant society paradigm and is therefore poorly understood by the dominant culture. For example, the recent *Native Title Act, 1993* passed by the Australian Government seeks to declare legal titles and interest in land (mining, tourism, etc.) without recognising the ongoing relationship between land title, resource use and conservation for Aboriginal people. It fails to address the claims of Aboriginal Australians in a realistic and sustainable way. Further conflicts will be generated without any framework for resolution and negotiation between Aboriginal and non-Aboriginal Australians.

Aboriginal rights relating to access, land and resources use and participation in 'caring for country' will be extinguished (without compensation in many cases) or left to the discretion of a vast array of Federal, State and Territory tribunals and courts. Overwhelmingly, these decisions will be made by non-Aborigines (assisted, in some cases by Aboriginal assessors) who will not necessarily have any background in bi-cultural procedures, dispute resolution or knowledge of the complex relationship between Aboriginal people and their land. Aborigines applying for native title, or defending interests in their land, will not be provided with additional resources outside the existing budgetary allocation for Aboriginal Affairs. Limited assistance will be available to some Aboriginal organisations 'approved' by the Federal Government.

When faced with these types of disadvantages, it is important to understand that the value of mediation to enhance citizen participation is not absolute. A process of full citizen participation in decision-making does little to address power imbalances between the various interest groups (unless, of course, the impartiality of the process is overtly threatened thus delivering it to the political realm)[4]. For example, some groups (such as community or indigenous groups) may not have the financial or expert (technical) resources to participate in a mediation process. This will be exacerbated if the mediation process is outside their social and cultural experiences. Formal legal adjudication and judicial review may also be of little assistance in protecting the fundamental social and cultural concerns of these groups.

The mediation of sustainable development policies, which involve public participation, needs to proceed from a basis of equality in bargaining power (including the legal recognition of rights) access to resources and expertise.

4. MEDIATION AND THE FORMULATION OF AUSTRALIAN POLICY ON SUSTAINABLE DEVELOPMENT

If proactive mediation is adopted, a number of different aspects of the decision-making process may be mediated. For example, it is possible to mediate *procedures*, *resources*, the *scope of issues* at stake as well as *specific outcomes*. Most of the literature on mediation has focused on very specific disputes relating to particular projects or relatively small numbers of issues. However, sustainable development may be a matter of high level government policy and the conventional approach to mediation is inappropriate. However, the mediation of specific disputes will be important in the *implementation* of sustainable development policy. I will focus on the federal Resource Assessment Commission (RAC) to illustrate how mediation could contribute to sustainable development policy in Australia. The legislation for the RAC required it to consider sustainable development principles in its decision. The RAC has been 'discontinued' but the legislation remains in force. Wider inquiries may be held under the federal *Environmental Assessment (Impact of Proposals) Act, 1974* in the future.

The *Resource Assessment Commission Act 1989* (Cth) established the Resource Assessment Commission (RAC) as a public inquiry body designed to provide advice to the Federal Government on options for natural resource use decisions within the criteria of ecologically sustainable development (Stewart *et al.*

---

4. That is, equal treatment of unequal parties by a mediation can only result in inequality: '*if mediators are to redress an imbalance of power they must do more than treat the parties equally, they must compensate for that imbalance by treating the parties unequally. The danger is that the mediators will thereby compromise their neutrality*' (Astor and Chinkin 1992).

1991, Resource Assessment Commission 1992). The RAC is responsible to the Prime Minister to whom '*it is to provide independent, comprehensive, informed and unbiased information*' on the issues referred to it (McColl l993). The RAC had three matters of national significance referred to it: Australia's forest and timber resources, the Kakadu Conservation Zone and the coastal zone. The unique nature of the RAC legislation and process make it worthy of retrospective review in the context of mediation and sustainable development strategies.

Section 8 of the *Resources Assessment Commission Act, 1989* sets out the matters to be addressed by the RAC in its inquiries:

(a) identify the resource with which the matter is concerned and the extent of that resource;

(b) identify the various uses that could be made of that resource;

(c) identify:
  (i) the environmental, cultural, social, industry, economic and other values of that resource or involved in those uses; and
  (ii) the implications for those values of those uses, including implications that are uncertain or long-term;

(d) assess the losses and benefits involved in the various alternative uses, or combinations of uses, of that resource, including:
  (i) losses and benefits of an unquantifiable nature; and
  (ii) losses and benefits that are uncertain or long-term;

(e) give consideration to any other aspect of the matter that it considers relevant.

The *Resources Assessment Commission Act 1989* set up a process which identified the need to dramatically improve our baseline information about the environment and to involve the public and relevant public and private organisations in the formulation of policy at an early stage. The process can be implemented earlier than a project specific environmental impact assessment and allows cumulative and sectoral impacts to be addressed. The need to evaluate social and cultural values is clearly required. They could also be included in the 'unquantifiable' and 'long-term' uses of resources which must be considered. Another advantage of this process is that the research and inquiry process are not under the control of private project proponents.

Forums, such as the RAC, which makes recommendations to the Federal Government on matters of national policy,[5] need mediation to develop innovative policy processes, inquiries and investigations. Although it is constitutionally appropriate that policy formulation ultimately rests with the elected government, many resource and environmental policies are so fraught with state, federal, industry and public conflicts that the traditional government decision-making process is ineffectual and an inquiry Report can carry significant weight: '*The RAC should be seen as a crucial element in the sustainable development debate. It will be at the forefront of policy development if it is able to embrace the full range of techniques for resolving resource and environmental conflicts through its processes*' (Boer et al. 1991a).

I was joint author of a Report titled *The Use of Mediation in the Resource Assessment Commission Inquiry Process* (Boer et al. 1991b) which was commissioned by the RAC to advise them on how mediation processes could be utilised in their inquiries. The Report looked at the use of mediation in developing focussed and participatory inquiry processes. The fruitful areas for mediation in environmental policy, by the RAC, were as follows:

- the identification of parties and interests;

- the terms of reference for inquiries (or the interpretation of them if the terms of reference are not subject to mediation);

- the issues (and the resources and priorities which will be given to them) which will be dealt with by the inquiry;

- inquiry procedures;

- research approaches and use of experts; and

- recommendations of inquiry (including policy options).

Figure 1 outlines how mediation could have been used in an inquiry process such as a RAC inquiry. The most significant recognition of social and cultural issues was in the inquiry into the Kakadu Conservation Zone. However, the RAC did not act on the proposals in our Report and there was a continual emphasis on economic and scientific factors usually considered in a more traditional inquiry process.

The most successful environmental inquiries, and policy making, that have ever been undertaken have involved comprehensive approaches to environmental and resource policies and the procedures have been the subject matter of intensive discussion and mediation. To a great extent, effective mediation in a policy process involves very intensive participation, good faith on the part of parties (particularly the persons responsible for convening the process) and the imagi-

---

5. Statutory inquiries to make recommendations to the Federal Government may also be established pursuant to the *Environmental Protection (Impact of Proposals) Act 1974 (Cth)*. However, this is on a specific project basis and, therefore, does not have the same scope to impact upon government policy and strategy as inquiries under the *Resource Assessment Commission Act 1989*. For this reason, this paper will focus upon the Resource Assessment Commission.

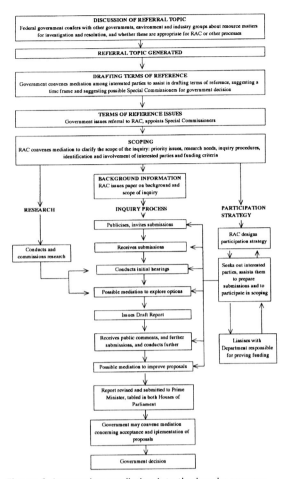

Figure 1. Integrating mediation into the inquiry process

native use of funding resources, access to expertise and research.

6. CONCLUSION

The concept of mediation as a broad participatory approach applicable to sustainable development is unconventional. I do not see processes such as public participation, mediation, environmental impact assessment and social impact assessment as mere techniques. They are attempts to improve the quality and adaptability of policies and decisions as well as to provide some new ethical directions for them. This does not always succeed and procedures such as public participation and mediation can quickly become bankrupt and support the status quo. However, the desire of the public to have ecological, social and cultural factors understood and valued has endured. Sustainable development is the most recent and comprehensive attempt to act on these concerns which have been around for a very long time.

References

Amy, D. (1990). Environmental Dispute Resolution: The Promise and Pitfalls. In: *Environmental Policy in the 1990s: Towards a New Agenda*, (eds Vigg, N. J. and Craft, M. E.), pp. 211. University of CQ Press, Rockhampton.

Astor, H. and Chinkin, C. M. (1992). *Dispute Resolution in Australia*. Butterworths, Sydney.

Boer, B. (1992). Implementing Sustainability. *Delhi Law Review*, **14**: 1-33.

Boer, B., Craig, D., Handmer, J. and Ross, H. (1991a). *The Potention Role of Mediation in the Resource Assessment Commission Inquiry Process*, Resource Assessment Commission Discussion Paper No. 1. Resource Assessment Commission, Canberra.

Boer, B., Craig, D., Handmer, J. and Ross, H. (1991b). The Use of Mediation in the Resource Assessment Commission Inquiry Process. Consultants' Report to the Resource Assessment Commission.

Cameron, J. (1989). Moving Towards Sustainable Economic Activity in Australia. A paper presented to the Australian Freedom From Hunger Campaign and Research Institute for Asia and the Pacific Seminar on Key to Survival — Sustainable Development, Sydney, July, 1989.

Craig, D. (1989). Sustainable Development and Indigenous People. A paper presented to the Second National Conference of the Environmental Institute of Australia, Environmental Practice & Sustainable Development, Melbourne, October 9-10, 1989.

Dixon, J. and Fallon, L. (1988). *The Concept of Sustainability: Origins, Extensions & Usefulness for Policy*. East-West Centre, Environmental and Policy Institute, Hawaii.

Kubasek, N. and Silverman, G. (1988). Environmental mediation. *American Business Law Journal*, **26**: 533-555.

McColl, G. (1993). Environmental Dispute Resolution. A paper presented to the 12th National Conference of the National Environmental Law Association, Canberra, 5-6 July, 1993.

O'Riordan, T. (1988). The Politics of Sustainability. In *Sustainable Environmental Management: Principles and Practice*, (ed Turner, R. K.), pp. 29-50. Belhaven Press, London.

Resource Assessment Commission (1992). *Methods for Analysing Development and Conservation Issues: The Resource Assessment Commission's Experience*. Research Paper No. 7, RAC, Research and Information Branch, Canberra..

Stewart, D. G., Mills, R. and Hamilton, C. (1991) *Perspectives on the RAC*. Resource Assessment Commission, RAC Occasional Publication No. 2, AGPS, Canberra..

Susskind, L. and Weinstein, A. (1980). Towards a theory of environmental dispute resolution. *Boston College Environmental Affairs Law Review*, **9**: 311-357.

World Commission on Environment and Development (1990). *Our Common Future*. Oxford University Press, Melbourne.

Chapter 12

Conflict Management: Achieving Practical Outcomes

Marg O'Donnell,
and Christine Nolan

*Alternative Dispute Resolution Division
Department of Justice and Attorney-General
Box 149, GPO Brisbane, QLD 4001
Australia*

Abstract

'Alternative dispute resolution' is a portmanteau expression that describes interventions into disputes by neutral third parties — they may be fact finders, mediators or arbitrators — as an alternative to court-based adjudication. Mediation is 'The intervention into a dispute or negotiation by an acceptable, impartial and neutral third party who has no authoritative decision making power, to assist disputing parties in voluntarily reaching their own mutually acceptable settlement of issues in dispute'. This paper addresses the application of alternative dispute resolution processes, and in particular mediation, in the resolution of environmental disputes.

Commonly, environmental disputes share a number of features. There is frequently a complex interplay of technical and scientific data, law, values, perceptions and emotions. These disputes involve matters of the public interest and are in the public domain. Commonly, multiple parties have an interest in the matter, and whilst some parties may not have standing before the court, they may have a significant capacity to generate continuing conflict. Environmental disputes are often concerned with non-renewable assets or fragile eco-systems. Once decisions are made, they have far-reaching and immutable consequences. Establishing the facts may be difficult, and the parties may be required to weigh up conflicting expert opinion about possible long-term detriments.

Case studies drawn from the practice of the Community Justice Program are discussed and the role of the technical or scientific expert in mediated resolution of environmental disputes is contrasted with the role of the expert in litigation.

Four possible roles for the expert in mediated settlements are identified:
- *the expert as consultant to the group i.e. all parties;*
- *the expert as consultant to one party;*
- *the expert as party;*
- *the expert as mediator.*

1. INTRODUCTION

There is a contemporary proverb that says *'The nearest thing to eternal life on earth is a law suit'*.

There is also Nils Christie, a Danish criminologist, who says that our conflicts are good for us, that they strengthen us, that we have much to learn from them and that we must think of conflicts as property and guard them jealously.

Christie (1977) says that in modern Western societies conflicts have been taken away from the parties directly involved and in the process have either disappeared or become someone else's property — usually lawyers', or the States'. This is a problem, he argues, because conflicts are potentially very valuable resources for us as individuals and as communities.

These two issues: the disadvantages of litigation, notably delay and expense, and the need for people to have control of and manage their own conflicts are themes which underpin the philosophy and operations of the work of the Alternative Dispute Resolution (ADR) Division in the Department of Justice and Attorney-General in Queensland. Most major conflicts involving issues of public interest and concern are never fully resolved. They move back and forth from an active to a passive phase. They often do not lend themselves to the strict confines of adversarial processes of court action or even of an inquisitorial Commission of Inquiry.

Of course, mediation and other forms of ADR are not a cure-all. Some matters are unable to be settled through one of these collaborative processes and will need to be adjudicated in open court. On other occasions, governments in particular, will need to exercise the power vested in them by the electorate and make executive decisions which are binding upon the community and for which they are ultimately held accountable. The suitability of a particular dispute for mediation will depend on the interplay of a variety of factors relating to the substance of the dispute, the nature of the parties to the dispute and the process of the dispute.

In this paper I will discuss the elements of environmental disputes, define mediation and report on some of the work of the Alternative Dispute Resolution Division, known by its service delivery name, the Community Justice Program. I will speak mostly about mediation of existing environmental conflicts with some illustrations from our case files. Mention will also be made of our work in conflict management and prevention and some illustration of projects involving the use of conflict resolution skills to defuse or minimise potentially damaging disputes.

The privilege and secrecy requirements of the *Dispute Resolution Centres Act 1990* (Sections 5.3 and 5.4) place constraints upon the Program's ability to discuss cases publicly. Those that are identified are matters which are or have been in the public domain or where participants have given permission for discussion.

2. ENVIRONMENTAL DISPUTES

Environmental disputes have features that differentiate them from other types of disputes. They are usually disputes over public or shared goods rather than private property. Elements of these disputes include:

1. There is a complex interplay of technical and scientific data, law, values, perceptions and emotions; witness the disputes over the damming of the Franklin River in Tasmania and ongoing disputes regarding the management and future of Fraser Island in Queensland.

2. They are often conflicts over important public choices. They affect the wider community and therefore the parties are unable to make private decisions.

3. They are often concerned with non-renewable assets or fragile ecosystems. Once decisions are made they may have far reaching and immutable consequences.

4. They often represent a clear and deep clash of cultural values. For example, there is the view on the one hand that the environment is fragile and interconnected, that it has intrinsic value and that there should be a risk averse attitude towards development. In contrast, some subscribe to the technocratic view which maintains that material growth is essential to human well-being and that this growth can only be achieved through mastery of the environment. There is a view that governments should exercise power and control over the environment for the greater good of the community versus the independent view that people who own land or resources should be able to make their own decisions about its future. Strong values almost always underlie environmental conflicts.

5. Environmental conflicts frequently involve all three levels of government — Local, State and Federal, and in some cases they stimulate intergovernmental conflict. The Franklin Dam case in Tasmania is a noted illustration of a bitter Federal/State environmental dispute.

6. They are often extremely difficult to manage, partly because there are multiple players often representing large and sometimes unwieldy reference groups. They also tend to be polycentric; the focus of action and issues tends to move, and resurface in another place or form.

7. Because of the requirement for technical and scientific input, there is a need to clarify and define the role of the expert. Sometimes parties claim they are experts. Sometimes experts claim they

are impartial. Management of expert information becomes crucial in the resolution of environmental disputes.

A brief diversion to define terms. Generally speaking, the phrase 'alternative dispute resolution' is a portmanteau expression that describes interventions into disputes by neutral third parties. They may be fact finders, mediators or arbitrators — as an alternative to court-based adjudication. Mediation is *'The intervention into a dispute or negotiation by an acceptable, impartial and neutral third party who has no authoritative decision making power, to assist disputing parties in voluntarily reaching their own mutually acceptable settlement of issues in dispute.'* (Moore 1989)

The work of the Community Justice Program centres on the provision of mediation services. The Program provides what is known as procedural mediation rather than substantive mediation, that is, our mediators do not advise on the content of a dispute but are proactive and authoritative in managing the dispute resolution process. At its best, mediation is a collaborative, problem solving, consensus building exercise.

At this point a case presentation will illustrate the type of matter which can lend itself well to the mediation process:

2.1 The Conondale Range Land Use Study

In late 1990 the Directors of Forestry (Department of Primary Industries) and National Parks (Department of Environment and Heritage) in Queensland set up a consultative committee to achieve the task of implementing the new Government's promise of an enlarged National Park in the Conondale Ranges in South-East Queensland. Issues surrounding the management of the Conondale Range area had been the subject of much debate and long term advocacy from both conservation groups and the timber industry.

The Consultative Committee comprised four stakeholder groups. These were three conservation representatives — Queensland Conservation Council (state), Sunshine Coast Environmental Council (regional) and Cooroora Forestry Association (local); three timber representatives — Queensland Timber Board (state), Boral Timbers (regional) and Grant Timbers Woodford (local); three Forestry representatives and three National Park representatives.

The Consultative Committee appointed a Zoning Working Group to examine all the data and to make recommendations to the Committee. Two representatives from each of the four groups sat on this body and they engaged the services of mediators from the Community Justice Program.

In December 1991 the Report and Recommendations of the Conondale Range Land Use Committee was released. It had been agreed to by all parties. It included recommendations regarding an increase in the range of park management practices, an agreement to produce a management plan, cessation of logging in certain key areas, and other matters.

Frances Stevens (1993) in her research on processes for resolving public interest land use disputes in Queensland interviewed a number of the parties some time after the completion of the Land Use Study. She found that all parties interviewed expressed positive feelings about the whole working party process. Commonly expressed positive aspects were:

(a) commitment to and ownership of the process;
(b) increased communication/better relationships between the parties involved;
(c) the process was based on a decision that had already been made; discussion could focus on the implementation of the decision.

It is worth noting the degree of satisfaction parties felt regarding the process itself as well as with the outcome. This is of course a strong element of the argument in favour of ADR, that is, that it is as important to satisfy procedural and emotional needs as it is to achieve substantive outcomes.

It is interesting to compare these responses with those of parties who gave evidence to the Commission of Inquiry into the Conservation Management and Use of Fraser Island and the Great Sandy Region conducted by Mr G E Fitzgerald QC in 1990–91. Some of the stakeholders here were the same actors as those who subsequently participated in the Conondale Land Use Study.

There have been many criticisms of the Final Report and the (Fraser Island) Inquiry Process (Stevens 1993). These criticisms centre very much around the processes of the inquiry. One was the observation that: *'The tendency for environmental disputes to be seen as scientific disputes able to be settled by impartial referees was based on an erroneous paradigm which regarded the conflict as a dispute to be resolved rather than an exercise in multi-objective resource planning.'* (Chenoweth 1992)

Another criticism was that the Inquiry did not provide an opportunity for interest groups to adequately state their views, especially as they were not able to meet with Commission staff. There were also concerns about the lack of transparency of analysis; the Commission was criticised for not explaining how it weighted all the conflicting perspectives to come to a conclusion.

2.2 Issues of Practical Consideration

While many identify the advantages of settling or at least scoping public land use disputes through the processes of ADR, there are many practical considerations which need to be taken into account for there to be a chance of conflict minimisation or management.

An obvious and early consideration is the process of *identifying and involving stakeholders*.

Bringing *all* of the relevant parties to the negotiation table is an essential feature of good conflict management work. It is often the most difficult to achieve, that is, both finding all representative interests and secondly, persuading them to meet. Identified lobby groups are sometimes easy to locate, but do not always represent all views.

In the Paddington Terraces dispute in Brisbane, another matter with which the Community Justice Program was associated, the Brisbane City Council, in an attempt to locate and involve all stakeholders, advertised a series of open public meetings with different interest groups scheduled for different evenings. Thus one meeting was for residents, one for local businesses, one for real estate agents and property owners, one for churches, welfare and the elderly and so on. All meetings were well attended, chaired by the Community Justice Program and had, among other tasks, the responsibility of selecting a set number of representatives to sit on a smaller ongoing working party. Each group completed the selection of representatives in their own idiosyncratic way and a working party of about fifteen, consisting of all above groups as well as City Council staff and aldermen and external town planning experts, met over a six month period.

In another case, in the beginning of the facilitation process over the Brisbane Landfill, the Mayor of Brisbane City wrote to all households within a defined radius of the landfill inviting them to attend a public meeting, encouraging them to consider who they might select for a small working party and what issues they would like to see dealt with by the working party.

The policy of the Community Justice Program regarding stakeholders is to be inclusive rather than exclusive. That is to say that if parties wish to attend and it is perceived that they have a voice or at worst can do damage to any agreement by sabotaging it from outside, then they should be included. Thus we have argued strongly in favour of inclusiveness with parties, often governments, who have wished to exclude certain groups, notably 'rat-bags', militant activists and other perceived trouble-makers, from the conference table.

Conversely, there can be difficulties when some key parties, especially community activists, refuse to attend. It is possible to facilitate discussion on these occasions between those other individuals and groups who will participate. However, the outcome achieved by such a process may well be simply a scoping exercise where parties use the opportunity to identify key issues in dispute, settle or agree on those they can and leave aside from this process those that require all parties' decision.

A further problem with representative working parties is the need for them to remain in touch with their own constituency. Once a working party becomes established and interactive, it can appear to outsiders that a cosy form of co-option and collusion has occurred. Those outside may feel that their views may be at best compromised, and at worst, betrayed by their representatives. It is our experience that this can occur very easily. Part of our process of facilitation in these multi-party and public issues disputes is to actively encourage and even facilitate meetings between representatives and their constituent groups.

The contribution of the technical or scientific expert to the settlement of land use disputes is, I imagine, of concern to many of you at this symposium.

Much has been written about problems associated with the role of expert evidence in litigation. At worst, scientific experts can appear to be prostituting themselves for particular interests; witness these comments by Professor Langbein of the University of Chicago (1985): 'At the American trial bar, those of us who serve as expert witnesses are known as saxophones'. This is a revealing term, as slang often is. The idea is that the lawyer plays the tune, manipulating the expert as though the expert were a musical instrument on which the lawyer sounds the desired notes.'

However, even in less extreme circumstances, certain problems can attach to the use of expert witnesses within a formal adversarial framework. His Honour Justice Geoff Davies of the Queensland Court of Appeal and the Litigation Reform Commission has said (1992): 'My own view is that a great deal of time and money is wasted in the engagement and calling of opposing experts who, in many cases, have themselves become advocates. Moreover it leaves the inexpert judge or jury with the task of deciding between what appear to be opposing views on often complex questions. I think that parties should not, as a general rule, be allowed to call expert evidence; that the only expert evidence given in a case should be that of court appointed expert.'

Turning to the role of the expert in mediated settlement, four possible roles can be identified.

- **The expert as consultant to a group** attempting to arrive at a resolution i.e. to all parties. In this case the expert has no decision making authority in his/her own right. All participants have equal access to the expert and the opportunities for collaborative problem solving utilising the expertise of the consultant are maximised.

- **The expert as consultant to one party.** This situation is most analogous to that of the litigation context. However whilst parties may introduce 'partisan' expert evidence into the mediation, the setting provides the potential for other parties to explore and question that evidence in a more flexible manner than the court process allows.
The ever present issues in mediation practice of

power imbalance and unequal access to information are relevant here. Mediators have a responsibility to encourage parties to inform themselves, to reality test information and proposals advanced during the mediation, and to terminate a mediation where a clear instance of bad faith bargaining is occurring.

- **The expert as party**. In this circumstance, the expert is one of the parties engaged in decision making but also possesses a superior knowledge or skills base. This has been a common scenario in the work of the Community Justice Program where a government department is attempting, at least in part, to share decision making with industry or community groups or other government departments and is also the major source of technical advice for the group.

  In some cases, the department has hosted or initiated the process, including engaging the mediators, and role strain can arise from these arrangements. They are not necessarily unworkable but role clarity and real demonstrations of good faith may be required for other parties to be willing to negotiate.

  Where a mediator has been effective in engaging disputants in a collaborative problem solving process, the potential is there for the expert, or experts where parties wish to bring competing technical opinions to the table, to play a more holistic and constructive role, than may occur in the litigation process.

- A fourth role for the expert may be that of **expert as mediator**.

I want to argue against the use of the expert as mediator, although I would not rule out a role for an expert to work in a team arrangement with a process facilitator. The extent to which the mediator requires familiarity with the subject matter of the dispute in order to maximise his/her effectiveness and efficiency varies from dispute to dispute.

Experience suggests, however, that there is significant value in having parties educate the mediators and consequently one another about the matters in dispute as part of the resolution process. A key tenet of successful mediation is to accept nothing at face value. The task of the mediator is to explore, and explore further, the bases of parties' positions, clarifying perceptions, testing assumptions, identifying the information on which positions have been adopted. In this respect, the 'ignorance' of the mediator can play a useful role.

The case against using content experts as mediators can be made as follows: experts inevitably in conceptualising the dispute will be influenced by their own technical and philosophical frameworks. They will filter and structure information and communication based on professional constructs. They may tend to rely upon their prior knowledge or upon suppositions about the nature of the dispute, rather than allowing parties to teach them and each other about the dispute. The temptation for the expert is to focus on the technical aspects of the dispute and to underestimate the significance of underlying values and perceptions held by parties.

A further temptation for the expert is that of leading parties to solutions. This is a problem because a sense of ownership of the outcomes of mediation is critical to its success particularly in the post mediation phase as parties grapple with often difficult implementation tasks.

Participants in the Conondale Ranges exercise discussed earlier specifically valued the 'content free' nature of the mediators' involvement in the resolution process. Ms Stevens (1993) reported that: *'The members of the Working Group found the contribution of the mediators invaluable to the smooth running of the zoning process. One of the perceived strengths by the participants of the mediators was that they were disinterested in the technicalities, having no knowledge of the subject matter under discussion. The mediators could therefore focus on managing the process.'*

3. POLICY DIALOGUES

An increasing amount of time is being spent by the Alternative Dispute Resolution Division on a relatively new and innovative application of mediation in the area of 'policy dialogues'. In these types of negotiated policy development or planning exercises a neutral third party facilitator assists various interests to formulate recommendations for consideration/adoption by decision makers. In the United States regulatory negotiation is common with various state and federal agencies including the Environmental Protection Agency and the United States Forest Service utilising the process as a matter of course (Adler 1989).

Consultations between government departments and communities and between government departments and other government departments have been facilitated by the Community Justice Program at the state and local levels and have concerned a wide variety of matters. Examples include: the review of policing of Torres Strait Islander communities; the Queensland Advisory Meeting to develop a strategy for Caring for Returned Human Remains and Burial Artefacts; law and order and licencing arrangements on Aboriginal communities; the Brisbane Landfill Local Advisory Committee; the Great Sandy Region Community Advisory Committee; the Sandy Creek Integrated Catchment Management Project; and various issues in the health portfolio.

4. CONCLUSION

There are various approaches and strategies which fall under the umbrella of ADR or, more appropriately, conflict management. They can include mediation of disputes where parties are locked into opposing positions, or facilitation of round table discussions which encourage purposeful planning, and the construction of useful future models of management and dispute system design. All of these different types of mediation efforts are predicated upon the belief that bringing parties together and skilfully assisting them to hear each other and to be heard will begin the process of problem solving and options generation.

It is also accepted that litigation is a necessary but by itself, insufficient, inconclusive and often unsatisfactory method for resolving problems that affect us all.

The role of good ADR practice in dispute *prevention* must not, also, be underestimated. The protagonists in the Conondales; the timber industry and the greens, the National Parks Service and the Forestry Service now have practice in cooperating and negotiating purposefully and with respect for each other. Future conflicts can be diverted or at best mapped by a well placed phone call. The people along the Terraces at Paddington have met and forged agreements and made concessions to and with each other. Disputes in this area will re-emerge but there is every reason to hope that in the future they will be managed by the parties themselves in a constructive way.

References

Adler, P. (1989). *Mediating Public Disputes*. Proceedings of an International Conference on Environmental Law. The Law Association for Asia and the Pacific and the National Law Association of Australia, Sydney, 14–18 June 1989.

Chenoweth, A. (1992). Fraser Island and the Great Sandy Region: Management Issues. *Australian Environmental Law News*, **3**: 69–78.

Christie, N. (1977). Conflicts as property. *British Journal of Criminology*, **17**: 1–15.

Davies, G.C. (1992). *The response of the courts and tribunals to the challenges of ADR*. First International Conference in Australia on Alternative Dispute Resolution, Sydney, pp. 47–92.

Langbein, J.H. (1985). The German advantage in civil procedure. *University of Chicago Law Review*, **52**: 823–835.

Moore, C.W. (1989). *The mediation process: practical strategies for resolving conflict*. Jossey-Bass, San Francisco.

Stevens, F. (1993). *Existing and Potential Processes for Resolving Public Interest Land Use Disputes in Queensland*. Honours Thesis, Faculty of Environmental Sciences, Griffith University.

Looking to the Future

Chapter 13

Changing Perceptions on Savanna Development

Malcolm Hadley[1]

Division of Ecological Sciences
UNESCO
7, place de Fontenoy
75352 PARIS 07 SP,
FRANCE

Abstract

Variability and the need for flexible responses to uncertain events have long been recognized as intrinsic characteristics of tropical savannas. More recent has been the bringing together of concepts and ideas that have different origins yet similar implications. Changing perceptions of various actors in savanna land use are reflected in the interactions between scientists of different disciplines, in terms for example of perceiving and reacting to complexity, working across multiple scales, and confronting different knowledge systems. At the same time as changes within the scientific community and its understanding of savanna systems, the last decade has seen changes in the private sector and its relations to tropical savanna resources and in community-based responsibilities and co-management arrangements. Questioning of the assumptions and tenets that underpin savanna land use and management have sharpened understanding and awareness of the differences between so-called equilibrium and non-equilibrium environments and of fundamental pathologies in ecosystem management, and have fuelled increasing interest in approaches to conflict resolution and consensus building. Emerging policy insights and suggestions include strengthening social technology at the local level, adopting administrative arrangements which are ecologically sensitive to the episodic variable nature of savanna systems, devolving responsibilities to local communities, and emphasizing community needs for regional rather than sectoral development. People should be seen as an integral part of savanna systems, with concomitant recognition that sustainable development is not an ecological problem, nor a social problem, nor an economic problem - but rather an integrated combination of all three. While there may well be no single vision of the future of tropical savannas, there is an emerging picture of the concepts and methods needed to deal with the generic nature of the problems of savanna development, and of the sorts of new partnerships and connections that are required for adaptive policy management.

1. The views expressed in this paper are those of the author, which are not necessarily shared by UNESCO

1. INTRODUCTION

Variability in time and space, and the need for flexible responses to uncertain events, have long been recognized as intrinsic characteristics of tropical savannas — for several decades by scientists who have studied tropical savanna, for centuries and millennia by the pastoralists and other resource users who have lived there.

For two decades, social scientists have been emphasizing that development in these regions should be rooted in flexibility, mobility and local level solutions. Biologists have described in great detail the various ways that animal populations respond to environmental variability.

If there has indeed been a fair understanding of these key savanna characteristics, for a number of years, what is perhaps recent is a convergence in the concepts, interpretations and analyses between natural and social sciences and a bringing together of ideas and thinking that have diverse origins yet similar implications (Scoones 1994). The growing recognition of the overwhelming failure of savanna development projects, and the reasons for that failure, is contributing to a rethinking and revamping in programme design and implementation and in the disciplinary mix and balance of research expertise.

The main focus of this contribution is to explore some of the implications of these changing perceptions of the various actors in savanna resource use, and to examine some of the trends in the development of new partnerships and connections among them.

2. CHANGING GLOBAL CONTEXTS AND SCENARIOS

One starting point for examining the possible future prospects for particular geo-ecological regions, such as tropical savannas, is provided by the evolving socio-economic and geo-political fabric of the world in which we live. The world is now very different from what it was twenty, ten or even five years ago, and this shift has been brought about in large measure because of changes in science and technology and because of technological innovations, particularly in the field of communications.

2.1 Living in a fractured global order

In terms of strategic planning considerations, Frederic Sagasti (1989) is one of those who has highlighted a whole cluster of changes that we face in the 1990s — a period that he has called the new fractured global order. Among the epoch-marking changes identified by Sagasti are the following.

- **Rapidly shifting political context:** movement towards a post-bipolar world, with a blurring of East-West differences; changing role of nation states, which are no longer able to exercise control over economic and social phenomena that take place in the world; tendency towards political pluralism, people's participation and democracy.

- **Changing patterns of world economic interdependence:** globalization of financial markets, an almost seamless web of transactions involving global trading of securities and global finance and capital movements; major changes in the content and direction of international trade (e.g. emergence of the Pacific as a major trading area); implications of national debt.

- **Implications of global competition:** fostering of new kinds of collaborative arrangements; changing role of non-governmental organizations; continuing pervasive role of transnational corporations.

- **Cultural transformations:** growing importance of religious and spiritual values; rise of fundamentalism as a main driving force of certain economic and political actions throughout the world.

- **Implications for developing countries:** increasing heterogeneous situation within and between countries (exacerbation of differences); need for tailor-made approaches that correspond to unique circumstances of the various developing countries; recognition of a mismatch in virtually every developing country between financial resources and social demands, and the challenge that this represents to governance.

Sagasti has put forward a series of propositions on the implications of these geo-political and socio-economic changes and technological advances for international co-operation in science and technology. The emergence of what might be called a fractured global order - an order that is global but not integrated, an order that puts all of us in contact but simultaneously maintains deep fissures between different groups of countries - constitutes a window of opportunity for experimentation and change.

There is need to re-think radically what we mean by development and progress. In the same way that development replaced the word progress during the last thirty or forty years, we now have to re-think the content and meaning of those two words, primarily because the implicit idea of catching up with the West is no longer viable. Perhaps the appropriate word to begin to use is that of empowerment, the idea that we should empower individuals to be able to choose what is best for them and to give them the capacity to decide their own destinies.

In facing the imperative of social and institutional innovation, we perhaps have to rid ourselves of quite a few sets of concepts, a kind of ideological baggage that we carry implicitly with us. The distinction between the public and private sector does not make too much sense at present. The distinction between market and planning doesn't make sense either. And there are many dichotomies that we have to go beyond and transcend, in order to be able to design institutions that are able to function better against the background of social changes in this newly emerging fractured global order. The role of what has been variously called civil society, and the third system is going to be much more important and the public and private distinction does not capture that richness.

Another proposition of Sagasti is that knowledge generation, dissemination and utilization - science and technology if you wish — will play an even more critical role in whatever we may be calling development or empowerment or progress in the coming years. If one were to redefine development or empowerment or progress at present, it might be defined almost exclusively in terms of the capacity to generate autonomously knowledge and the capacity to use it. This is the real distinction between those countries, those individuals that are empowered to act as full human beings and those that are not.

2.2 Towards a bifurcated world?

The divergences of Sagasti's fractured global order find echo in the bifurcated world described by Thomas Fraser Homer-Dixon, of the University of Toronto's Peace and Conflict Studies Program, who has argued the case against separating social and political policies from the physical world of climate, public health and the environment (Homer-Dixon 1991).

For too long, we have been prisoners of 'social-social' theory, which assumes that there are only social excuses for social and political changes, rather than natural causes too. The social-social mentality emerged with the Industrial Revolution, which separated us from nature. But nature is coming back with a vengeance, tied to population growth and an increasing bifurcated world.

Homer-Dixon (quoting from Kaplan 1994) has drawn an analogy with a stretch limousine in the potholed streets of New York, where homeless beggars live. Inside the limousine are the air-conditioned post-industrial regions of North America, Europe, the emerging Pacific rim and a few other places, with their trade summitry and computer-information highways. Outside is the rest of humankind, going in a completely different direction.

In Homer-Dixon's bifurcated world, part of the globe is inhabited by Hegel's and Fukuyama's Last Man — healthy, well fed and pampered by technology. The other larger part is inhabited by Hobbes' First Man, condemned to a life that is poor, nasty, brutish and short. Although both parts will be threatened by environmental stress, the Last Man will be able to master it, the First Man will not.

How does the world of the tropical savannas fit in such a scenario?

- The deceptiveness of political maps in a world of increasing demographic, environmental and societal stress worldwide, in which criminal anarchy emerges as the real 'strategic' danger.

- The meaninglessness of borders dividing countries in several parts of the tropical savanna world, exacerbated by the colonially erected borders that were and are at cross purposes with demography and topography.

- The withering away of central governments and their feeble ability to function and implement even marginal improvements.

- The emergence of culture and tribe as the real borders and the rise of tribal and regional domains.

- The unchecked spread of disease and the wall of disease that threatens to separate Africa and other parts of the Third World from other regions of the planet in the twenty-first century.

- The growing pervasiveness of war and the anarchic implosion of criminal violence.

- The draining of people from the countryside into dense slums in coastal areas.

- The notions that humans have challenged nature far beyond its limits, that the environment and its natural resources (e.g. water, the most important fluid of the coming decades) will be the core national-security and foreign policy challenge of the early twenty-first century.

2.3 Trends in the world's savannas

This somewhat bleak picture of the world, Kaplan's (1994) 'Coming anarchy', in turn finds a reflection of some of the trends in savanna lands over the last couple of decades, underlined in several recent reviews and assessments (e.g. Behnke 1994, Hadley 1993, Scoones 1994, Young and Solbrig 1992, 1993).

Progressive marginalization has occurred in many savanna societies, in political, cultural and economic terms. Although linked to national economies and through them to world markets, many savanna people have little control over their own destiny. One result is a dramatic loss of cultural identity in many savanna areas. Outside pressures for modernization and the introduction of new policies frequently undervalue and even ignore existing cultural systems.

National and international policies are often formulated without considering their implications for the people inhabiting the world's savannas. Yet the influence of these policies on savanna land use can be dramatic. Land alienation for agricultural purposes is

one problem. Alienation by placing land in national parks and under other forms of conservation title without adequate compensation to traditional land-users is another source of disruption and discontent.

In some areas there is also a bias towards the introduction of schemes that privatize parts of the savanna. For example, Botswana's Tribal Grazing Lands Policy, which aimed to prevent overgrazing by settling pastoralists with large herds onto ranches, has been an abject failure (Pearce 1993). Overgrazing (the problem it was supposed to reduce) and social tension increased. When the policy was introduced it was assumed that wealthier pastoralists would move their livestock onto private leasehold ranches and give up their community grazing rights. Most, however, did not give up their traditional grazing rights and, instead, began to use both resources — exacerbating the problem that the policy was supposed to resolve.

Trade policies and subsidies seem to have similar effects in distorting information and reducing opportunities (Young and Solbrig 1993, page 342). Tariff policies can be used by countries to restrict imports and transfer wealth from rural to urban populations. In Zimbabwe, for example, import restrictions and pricing policies effectively tax savanna land-users in order to subsidize the manufacturing industry and city dwellers. In many industrialized countries, subsidized agricultural products and trade restrictions protect farmers from having to compete with savanna land-users, with the result that prices are lower than they otherwise would be and opportunities for savanna development are reduced significantly.

Tariff barriers and price support within industrialized countries have also created opportunities for the European Community to severely distort production. Case studies from Botswana and Zimbabwe both indicate that the immense profits that a few can make under the preferential trading agreements like the Lomé agreement tend to reduce prospects for a transition to sustainable savanna land use. In Botswana, for example, in order to meet herd health standards set by the European Community, large veterinary fences have been built across much of the country. This has enabled a few local ranchers to profit from access to European Community markets, but the cost has been the exclusion of wildlife and lost opportunities to develop game ranching as wildebeest, gazelle and other migrating species have less opportunity to pass through much of the country.

Development assistance can also have devastating effects on savanna land-users. In one example, in Tanzania, bilateral aid coupled with government input subsidies have caused cropping to expand onto *muhajega* land which plays a keystone role in the grazing system maintained by the Barabaig pastoralists (Lane and Scoones 1993). The Barabaig do not agree that a national desire for self-sufficiency in wheat production justifies the destruction, without compensation, of their grazing system.

Policies such as these, underpinned by western science and technologies on the one hand and conflict and civil strife in several areas on the other, has had a devastating effect in non-equilibrium environments characterized by high unpredictability and immense variability. Contributing to the failure of development projects over the last 20–30 years in many savanna areas has been the philosophy of imposing a linear logic on a non-linear world, through blueprint technical solutions that so-often ignore the important contextual issues of policy, history and culture and the intrinsic characteristics of savanna environments (Scoones 1994). A new vision for savanna systems calls for new roles for different actors, new professionals, new partnerships and new connections.

2.4 Scientific bridges

The role of science and scientists in relation to tropical savannas can be viewed at different levels. Improving understanding of how savanna systems function and respond to stresses of various kinds, and suggesting technical ways of improving savanna productivity and stability of resource use in particular savanna localities, are two of the traditional functions of savanna science. Yet the record of practical accomplishments is very questionable, and the question could fairly be posed on the extent to which the scientific community is guilty of being experts in tunnel vision and degrading technologies, strategically mediocre to wit.

2.5 Encouraging interactions

In savanna regions, as elsewhere, attempts to promote interdisciplinary approaches to applied research on complex land use problems have shown their worth, if given a chance. But the constraints encountered in putting these approaches into practice have been many and deep-seated. Some of these constraints are of a scientific character or internal to the scientific community. They concern, for example, the quality and quantity of existing information and local manpower, the behavioural and psychological characteristics of individual scientists, career considerations, and questions of method and approach. Other constraints are of a more administrative, institutional, political and financial nature (di Castri and Hadley 1986).

The common precursor and deep-seated root of many of the obstacles to interdisciplinary effort (institutional, financial, behavioural, methodological) is the prevailing system of education, based as it is on disciplines. With a few notable exceptions, educational institutions, throughout the world, at all levels, are organized along nineteenth-century disciplinary lines, and perpetuate traditional disciplinary views of the structure of knowledge. Various disciplinary fields have their own sense of corporatism, whose members favour - consciously or not

— the approaches, paradigms and information networks of their own particular discipline and profession. A trait encapsulated in George Bernard Shaw's remark (quoted from Rittel and Webber 1973) that 'every profession is a conspiracy against the laity'.

The need to break out of the vicious circles spawned by disciplinary-based educational systems calls for shifts in the ways that scientists from different disciplinary backgrounds interact with each other, and the overall paradigms that they use for working in multi-disciplinary teams. Francesco di Castri has used an analogy with a tree's roots to picture the sorts of relations that need to be encouraged between scientists with different disciplinary specialities. Just as a tree may have a tap root and lateral roots, with complementary functions, might it not be possible for at least some of the individual members of the scientific community in addition to having one main disciplinary root, also to develop lateral roots that interact with the roots of other plants for mutual cross-fertilization (di Castri 1992).

Another botanical analogy of di Castri (1981), subsequently developed by Webb (1990), conceives ecology - which is both a discipline and an approach — as a bunch of roots formed by contributions from many disciplines, with a dynamic core of unified knowledge (Figure 1). The resulting plant (a handy geophyte) is a shrub or vine of indefinite growth. The fruits have an organic rather than technological flavour, and include both utilitarian and intangible values that are not easily measurable. They often represent ends (e.g. episodes of personal fulfilment, scientifically based insights with human meaning), as well as guides to holistic management and regional planning. The contrast is with the tree of knowledge, as exemplified by the biological science, which qualifies as part of orthodox and restricted science in the Baconian-Newtonian mould insofar as it is able to quantify data and become reductionist. Each branch bears a variety of technological fruits. Their value can generally be measured, e.g. by efficiency as in engineering, or by optimization as in economics. Some, as in agriculture or public health, merge with cultural, personal values, and become less measurable.

With these botanical analogies in mind, it is perhaps instructive to examine whether progress is being made in savanna regions on seeking links between disciplines and encouraging disciplinary contributions to a core of unified knowledge. Three sets of examples are addressed in the following few paragraphs, which deal respectively with questions of perceiving and reacting to complexity, working across different scales, and making links between different knowledge systems.

2.6 Perceiving and reacting to complexity

One consequence of professional training along disciplinary tracks that follow those established in 19th century academia is that observation that is needed and natural to pastoralists and other savanna users has often been trained out of professionals whose fields of enquiry and curiosity are narrowed along physical and disciplinary lines. One consequence is that complexity and diversity in agricultural systems are underperceived and accordingly undervalued (Chambers 1983, 1990). Non- or biased perceptions are manifest in many ways: sites for

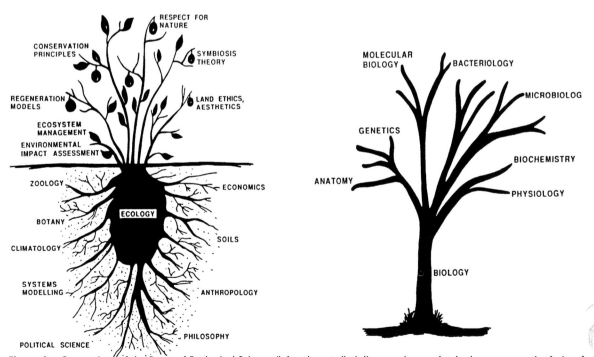

Figure 1. Comparison of the Roots of Ecological Science (left, relevant disciplines are incompletely shown, as are the fruits of the plant) and the Tree of Knowledge, as exemplified by biological science (*right*, extent of diversification indicative only). *Source*: Webb (1990)

agricultural research and trials tend to screen out variability in soils, topography and other conditions; field visits by scientists are vulnerable to the biases of 'rural development tourism' — spatial, project, person, seasonal, professional and diplomatic; short-time horizons in the on-farm situation are much more characteristic of the non-farming professional than of the farmer; perceptions by outsiders of the livelihood strategies of rural people tend to be simplified and stereotyped.

Chambers' (1990) phenomenon of 'Micro-environments Unobserved' is one reflection of this screening out of much the complexity and diversity of agricultural systems. Micro-environments are more separate and distinct than those created by such agricultural practices as ploughing, irrigation, mixed cropping, grazing and browsing, and differ markedly from their surroundings, in presenting sharp gradients or contrasts in physical conditions internally and/or externally. Micro-environments can be isolated, or contiguous and repetitive, and natural or made by people or domestic animals. They include home gardens, vegetable and horticultural patches, river banks and riverine strips, levees and natural terraces, valley bottoms, wet and dry watercourses, alluvial pans, artificial terraces, silt trap fields, raised fields and ditches or ponds, water harvesting in its many forms, hedges and windbreaks, clumps, groves or lines of trees or bushes, pockets of fertile soil (termitaria, former livestock pens, etc.), sheltered corners or strips (by aspect of slope, configuration, etc.), plots protected from livestock, flood recession zones, small flood plains, springs and patches of high groundwater and seepage, strips and pockets of impeded drainage, lake basins, ponds (including fishponds), animal wallows, etc.

Micro-environments such as these have a number of important properties and functions. They tend to differ from their more uniform surroundings and fulfil specialized functions. They may serve to concentrate soils, water and nutrients in ways that can be better tapped and used by farmers. They may provide many benefits to rural households and provide reserves and fallbacks to meet contingencies, lean seasons and bad years. They play a vital part in innovation, experimentation and adaptation and in many situations they have been developed as a form of intensification linked with increasing population density. In future, as rural populations continue to increase in many regions, micro-environments will become even more important for the livelihoods of poor farming households.

In terms of agricultural research and support, micro-environments demand new perceptions and quiet professional revolutions. Chambers (1990) concludes that perceptual shifts and professional revolutions will start not with the lecturer but with the farm family, not just in the classroom but in the field too, not on the research station but in the micro-environments themselves. They will entail not simplifying and standardizing but enabling farm families to complicate and diversify. The 1990s will show whether non-farming professionals can make that revolution and usefully meet that challenge, or whether it will be largely unassisted that farmers continue to experiment, innovate, develop and manage on their own.

Certainly, the last two decades have seen a shift in perceptions on development paradigms (Figure 2) — with increasing questioning of the top-down, centre-outwards approach to the transfer of technology and the evolution of a new family of complementary participatory approaches to agricultural research and extension, variously described as farmer-back-to-farmer, farmer participatory research, farmer first, etc. Often these two paradigms have been presented as opposing and antagonistic — cf. del Bono's (1988) *I Am Right, You Are Wrong* — rather than potentially complementing and mutually enriching. More recently, there are indications of new ways of linking the two. One example is provided by frameworks within the programme on Tropical Soil Biology and Fertility (TSBF), which has entailed participation of agricultural scientists, ecologists and most recently social scientists in seeking ways of combining on-station and on-farm research, and of articulating system and process-focused research (Figure 3).

3. WORKING ACROSS MULTIPLE SCALES

A wide range of space and time scales characterize the processes and phenomena which interact to shape environmental condition and trends. Important perspectives of environmental space and time include the role of terrestrial and extraterrestrial factors in shaping climatic change, insights to be gained from the prehistorical record, relations between disturbance and biotic responses, episodic events and large-scale phenomena, cumulative impacts, fast-slow processes and memory reservoirs. Scales in physical, chemical and biological phenomena have parallels in human driving forces, societal relations and decision-making

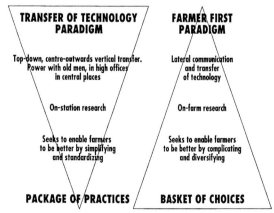

Figure 2. Contrasting development paradigms. Based on Chambers (1990).

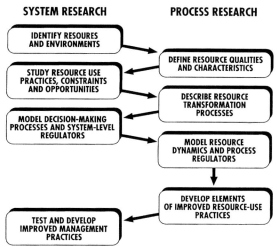

Figure 3. Iterative feedback of information in soil management research at the agroecosystem (on-farm) and process (on-farm and on-station) levels to develop improved soil management practices. An evolving framework from the Tropical Soil Biology and Fertility (TSBF) Programme. *Source*: Swift *et al.* (1994)

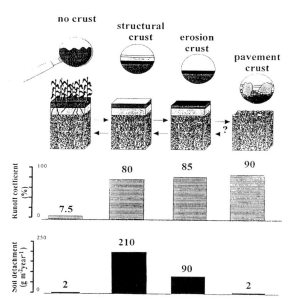

Figure 4. Time-sequence of surface crusting and removal by erosion of the top layers of sandy soils in the Sahelian zone; consequences on runoff production and soil detachment. *Source*: Valentin (1994) and other references cited therein.

processes, and environmental scales of space and time thus have perceptual as well as physical ('objective') dimensions. Scale is clearly more than just size and dimension, and there is a growing body of examples on how zooming along and across hierarchical scales can help in seeking explanation ('how') and significance ('why'), and in revealing emergent properties. Scaling can also act as a motor for new approaches to scientific co-operation (di Castri and Hadley 1988).

Interest in scaling issues has grown immensely over the last decade, within the scientific community. The International Geosphere-Biosphere Programme (IGBP) has given a real boost to this process, with recent meetings held in Ensenada, Mexico (January 1993) and in Woods Hole, USA (May 1994) serving to mark the progress that has been made in moving towards 'Integrating Earth System Science', and to fulfil the earlier vision of 'a new kind of science, genuinely innovative, interdisciplinary and international, courageously comprehensive, resolutely integrated, and sharply focused on the scientific issues vital to meeting humanity's needs' (J. Perry, cited in *IGBP Global Change Newsletter* 13, March 1993, page 2).

In terms of work on terrestrial systems, consortia of researchers (from different disciplines and countries) are developing projects which are attracting substantial levels of funding (e.g. from the European Community). Large-scale gradient and transect work is underway or planned (e.g. in West Africa, southern Africa, northern Australia, high latitudes in North America and Europe). Large-scale co-ordinated experiments are taking shape. In short, the impression is that researchers from a range of bio-physical disciplines are at last starting to work in a concerted "bio-science' way, at multiple scales, as has long characterized the oceanographic, atmospheric and geophysical communities.

In savanna regions, an example of the implications of linking research at different scales is provided by studies in West Africa, entailing intimate understanding of processes at the soil particle level and the implications to changes in soil infiltration and runoff at landscape and regional scales. Analysis of the processes and factors involved in surface crusting has led Casenave and Valentin (1989, 1992) to propose a morphogenetically based classification of the main surface crusts in the arid and semi-arid areas of West Africa. Nine main types of crusts have been differentiated according to number and texture of microlayers as well as the structure of the outcropping microlayer. The crusting processes have marked effects on run-off and soil detachment (Figure 4), emphasizing the need for experimental programmes aimed at improved process-based understanding, incorporating different scales and integrating process-scale issues (Valentin 1994).

A somewhat different aspect of the scaling question concerns comparison of phenomena and process at the same generalized unit of scale (such as stand or patch) in two or more ecosystem types. One of the long standing divisions of resource managers and research scientists is according to environmental media and system types. One broad grouping is between scientists and resource managers concerned with marine and terrestrial ecosystems and points of contact and collaboration between those groups have tended to be rare. On land, one set of groupings of scientists and managers is along ecosystem or physiographic lines, as

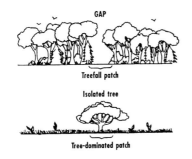

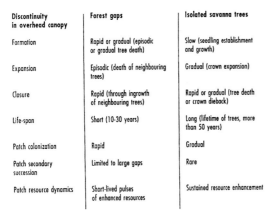

Figure 5. Comparison of the dynamics of gaps and patches formed by treefalls in forests and of isolated trees and tree-dominated patches formed by the growth of trees in tropical savannas. Source: Belsky and Canham (1994)

reflected in a wide range of professional societies and periodicals catering for those interested, for example, in grasslands, forests, mountains, arid lands, wetlands, and so on. Such are the differing interests and constituencies that opportunities for co-operation may be not seized or indeed not even appreciated. Yet cross-fertilization between scientists working in different systems would appear to hold considerable promise.

The application of patch dynamics in forest and savanna systems represents a case in point (Belsky and Canham 1994). Researchers have documented remarkable similarities in the processes of canopy-gap formation and closure across a wide range of forest ecosystems. In the paradigm that has emerged from these studies, an intact forest canopy is viewed as a closed matrix in which are embedded a population of gaps that can be described in the terms of traditional organism-level demography: birth (gap formation), growth (usually negative), and death (eventual gap closure). Although this model of patch dynamics has proven to be a powerful tool for organizing understanding of forest dynamics, it has seldom been applied to other terrestrial ecosystems.

In proposing that tropical savannas represent another important ecosystem in which the model of patch dynamics may profitably be applied, Belsky and Canham (1994) have compared savanna trees and forest gaps and have identified striking similarities between the two community types (Figure 5). They have also suggested that continued comparison of patch dynamics in forests and savannas may produce important generalities about the interplay of patch and community dynamics in terrestrial systems.

4. Bringing Together Different Knowledge Systems

Increasing dissatisfaction with the failure of western scientific knowledge and technology to contribute effectively to the development of sustainable resource use systems in the tropics is among the factors that have fuelled a growing interest in the traditional knowledge of farmers and other resource users. The underlying rationale can be simply stated. Local pastoralists, cultivators, fisherfolk and other resource users often have a profound knowledge of their highly varied environments which could be better tapped in assessing the potential use of biological materials and the development opportunities of their homelands. Much of this knowledge is unrecorded and unexploited, and every year part of this knowledge is being lost with the transformation of ecosystems and local cultures. Recording and applying this traditional ecological knowledge provides one approach to making more effective use of the biological wealth of the planet, particularly in terms of biological diversity and genetic richness and their use as a starting point for strategies of integrated conservation and sustainable development. There may also be considerable scope in diffusing information on techniques and practices refined over generations in one part of a particular ecological region, and testing and adapting them in other localities.

Within such a context, increasing appreciation among scientists of local ecological knowledge over the last 20 years has led to the launch of several international journals and newsletters, including *Etnoecología*, *Journal of Ethnobiology* and *Indigenous Knowledge and Development Monitor*. This trend has also inspired the founding of the International Society of Ethnobiology, which organized the first International Congress of Ethnobiologists in Belem (Brazil) in 1988 and subsequent meetings in China, Mexico and India. It has also led to a growing appreciation among development workers and policy-makers of the contribution that indigenous knowledge can make to development projects, in such fields as integrated approaches to natural resource management, holistic medicine, social control mechanisms for common property resources, using sacred groves for ecosystem rehabilitation purposes, combining local classificatory schemes with modern survey technologies. Two examples follow.

In many parts of Africa and Asia which have been largely modified by human activities, intact patches of land remain in the form of sacred groves, which house the gods of village communities and which are therefore taboo to human interference. Among other values, sacred groves

contain a store of biological diversity, which can perhaps be tapped in efforts to repatriate surrounding areas of degraded land. Such is the underlying motivation of a project entitled 'Co-operative Integrated Project on Savanna Ecosystems of Ghana' (CIPSEG), supported by Germany through funds-in-trust arrangements with UNESCO and Ghana. Field reconnaissances in late 1992 led to the identification of three sacred groves (out of forty) as an initial focus for project activities: the Tali (or Leopard) grove in the Tolon/Kumbungu District; the Malshegu (or 'Python' grove) in the West Dacomba/Tamale District; and the Yaroogo grove in the Saveluga/Nanton District. The project entered its operational phase in early 1993, with a first group training activity being held in Tamale in northern Ghana in July 1993, culminating in a workshop on strategies for mobilizing and motivating communities towards savanna ecosystem management. Over the training period, 150 persons participated in lectures and field activities, with training and demonstration in such fields as seedling nursery management and tree seedling planting techniques developed by the Forestry Research Institute of Ghana for use in the environmental stressed environments of the Guinean savanna of northern Ghana. Over 500 tree seedlings (including *Acacia senegalensis*, *Adansonia digitata* and *Ceiba pentandra*) were planted on a two-hectare highly-degraded transition zone around the Malshegu sacred grove.

Indigenous knowledge can also contribute to more useful and more cost-effective ways of describing the natural resources (soil, water, vegetation) of a given region, by complementing — not replacing — conventional survey techniques. One example from the Senegal River Valley (Tabor and Hutchinson 1994) concerns the integration of the classification systems used by local resource managers with remote sensing and geographic information systems. In the drought-prone Senegal River Valley, three riparian landscapes are recognized by local farmers. One is flooded by the Senegal during one period of the year and the other two by local rains in another season. Although each landscape is composed of the same soils, they are distinguished by the frequency, period and time of year they are flooded. As a result, each soil cannot be recognized in a conventional soil survey because distinctions are based on value and management criteria rather than physical properties. Yet the insights from local knowledge can be used to enhance the efficiency of conventional information-gathering. Using indigenous knowledge ensures that the most important factors determining resource value and management practices will be captured during the survey. In addition, use of local terminology ensures that the information generated can be more readily disseminated.

5. Private Sector Roles and Linkages

The last decade has seen important changes in the relations of the private sector to environmental concerns. There has been increasing recognition by the private sector of the commercial opportunities opened by environmentally friendly 'green' products. A whole welter of new initiatives have been suggested or launched aimed at promoting best practice standards for environmental management systems, environmental impact assessment and environmental auditing (e.g. Vaitilingam 1993).

Within such a generalized context, private sector enterprises have diverse functions in relation to tropical savanna resources. Companies could be classified in four large groups: (a) harvesters/transformers/producers of tropical resources; (b) researchers on tropical savanna goods and services; (c) financiers of private and public investments; and (d) trade and transport enterprises.

In each of these domains, there would seem to be considerable scope for seeking new partnerships with the private sector. Traditionally, there has been a polarization of perceptions between the community of conservationists, research workers and local people on the one hand, and the development community on the other (Muul 1993). The former have tended to accuse the government and private sector interests of collusion in promoting unsustainable land management practices, a process fuelled by low royalty rates, tax incentives and loan concessions linked to bush clearance, etc. The development community have in turn tended to be scornful of the naivety if not ill-intention of those advocating a less wasteful, a more conserving, approach to the use of the resources of the tropical savannas.

Recent years have, however, seen a shift in perceptions, at least among some of those concerned with tropical savannas and their use. Faced with widespread public dissatisfaction with past conservation and development strategies, some concerned individuals from among both the community of developers and the community of conservationists are beginning to reach across the perceptual chasm that has separated them. A new 'community' is forming in several parts of the tropics comprised of these concerned individuals. Working co-operatively, instead of at odds as in the past, these individuals as a community believe that they can be mutually supportive in seeking to combine conservation and long-term economic development.

Broadening the type and style of private sector involvement could take advantage of likely increasing international demand for public environmental goods (e.g. tourism, biological diversity, stable climate). This is one interesting trend, which perhaps could become an important focal point for future private sector co-operation and diversification. Among the sorts of entrepreneurial diversification that might be encouraged are: arrangements with the tourist industry (including airlines, tour operators and hoteliers), for the development of nature tourism ('tourism of discovery'); with

firms involved in beef production, entailing a crash programme in building-up breeding stocks of improved breeds of native cattle; and with horticulture, food processing, transport and retail outlets, for the improvement and marketing of tropical fruits.

5.1 Community responsibilities and co-management arrangements

Those who take an interest in the problems of natural resources are gaining a better understanding of the effect of institutional arrangements in enhancing or detracting from effective governance and management. In particular, we are learning more about institutional arrangements designed or evolved by resource users themselves. In all parts of the world, there is growing interest in encouraging community responsibility as one element of co-management resource strategies, in preserving and enhancing the diversity of management regimes, and in establishing partnerships in management responsibilities. Among these, community management is advocated not as an exclusive responsibility, but as one element in a co-management arrangement involving public and private interests to manage resources, which may be under private, public or communal ownership. The justification for community management can be found in at least four levels (Renard 1991).

First, community-based management promotes democracy and equity because it gives members of the community a greater opportunity to share in the decisions about how resources are used, and thereby a greater share in the benefits that are gained from their use. Priorities are no longer predetermined from the outside by bureaucracies or in boardrooms far removed from the everyday concerns of the users. Rather, they are developed from within by those whose livelihoods are directly affected by those choices. Similarly, it enhances the opportunities to increase the local benefits of resource use because the means of production are more likely to be smaller in scale and owned by the users themselves.

Second, community-based management is economically and technically efficient. Users have more clearly defined responsibilities for their decisions and actions and can provide a wide variety and considerable quantity of local resources, land, skills, technology, labour, capital, knowledge, and infrastructure to implement them. In particular, local and traditional knowledge and resource monitoring by community members can provide significant information to planning and development agencies on the characteristics of a resource. Local responsibility also decreases the need for costly outside enforcement, which many governments cannot afford.

Third, community management is effective because it is adaptive and responsive to variation in local social and environmental conditions and changes in those conditions. Often it is the failure of centralized strategies to accommodate the local socio-cultural conditions, not the resource conditions, that leads to the failure of the strategy. Furthermore, resource users are constantly aware of the condition of the resources upon which they depend, and they can be quick to respond and adapt to changes in the condition of those resources.

Fourth, local community control brings a measure of stability and commitment to management that a centralized government approach cannot duplicate. Government decision-making usually operates over a relatively short-term time horizon and is often met with resistance on the ground. On the other hand, people will show more commitment to decisions they have made themselves based on their own priorities. These priorities should reflect the objectives of long-term socio-cultural and resource sustainability, which are clearly in their best interest to ensure.

Renard (1991) goes on to propose five primary directions for change and action on the institutional front: strengthening community-based organizations; defining the role of non-governmental organizations; reforming governmental institutions and their operations; revising the role of multilateral, bilateral and donor agencies; and designing new approaches to training and education. Such orientations and approaches seek to contribute to a process that promotes social equity, respects popular needs and wisdom, and maintains cultural integrity and sovereignty. It is scarcely surprising that these orientations meet with a certain opposition where there is fear of such a process releasing new talents and redistributing power and responsibilities. Institutional change will be slow, but it would appear indispensable to achieve the goals of resource management and community development. Several examples can be drawn from tropical savanna regions.

One such example is Zimbabwe's CAMPFIRE initiative (Murphree and Cumming 1993, Young and Solbrig 1992, page 37). In Zimbabwe, concern about failure of land privatization schemes and the top-down approach to resource development, triggered the development of the Communal Areas Management Programme for Indigenous Resources (CAMPFIRE). This programme devolves management responsibility for, and benefits from, natural resource management, particularly wildlife, to local communities. Implemented in January 1989, the programme assumes that:

- Wildlife should be promoted as an economic form of sustainable resource use to enhance rural productivity;
- Local communities and landowners are more effectively motivated to conserve wildlife when this is of direct economic benefit to them;
- Sustainable exploitation of the resource requires cause-and-effect relationships linking good husbandry with benefit;
- Proprietorship must include the authority to

decide whether to use wildlife at all, to determine the mode and extent of its use, and the right to benefit fully from its exploitation;

- In communal contexts the unit of proprietorship, with rights of inclusion and exclusion, should be as small as management considerations permit, allowing conformity to management regimes and enforcement through collective but informal pressure (The unit of proprietorship should be the unit of management and the unit of benefit).

Whilst the CAMPFIRE initiative still faces a number of implementational problems, including bureaucratic resistance in certain circles and an inadequate legislative base, and is still too young to be confidently cited as a success story, it has nevertheless already had a dramatic impact on the financial base of operations for certain district councils (Murphree and Cumming 1993). Some councils are already reporting a sharp increase in incomes. In certain wards, producers now regard wildlife as an asset rather than a liability, and 'free rider' exploitation in the form of poaching has diminished. Communities have become more assertive over their claims to the proprietorship of all natural resources and have begun to make their own land-use plans to exploit and conserve the range of resources available. Collectively more confident about their own prerogatives and abilities, they are more conservative about the allocation of their land and more aware of the linkages that bind them to regional and national structures of resource exploitation.

Another example is the landcare groups in Australia - largely autonomous local groups of people, mainly land users in rural areas, whose objectives are to deal with problems of land degradation and to develop more sustainable farming systems and land-management practices. Although the landcare movement had its origins earlier, it was given a significant boost by the decisions of the Australian government in 1989 to designate the decade beginning in 1990 as 'The Decade of Landcare' and to provide over $320 million for landcare and related tree planting and remnant vegetation conservation programmes to the year 2000 (Hawke 1989, cited from Black and Reeve 1993). A major emphasis in landcare is the preparation of farm and regional plans which describe land resources, identify biophysical constraints, and help individuals and communities to move toward sustainable land use. Within such a context, the number of landcare groups has grown rapidly, from about 150 in 1987, to about 225 in 1988, 340 in 1989, 580 in 1990, over 900 in 1991 and about 1300 in 1992. Attitudinal variables appear especially important in exploring or predicting landcare group membership in the movement's early stages, though situational variables become increasingly important as the proportion of farmers participating in the scheme increases (Black and Reeve 1993).

Yet other examples can be drawn from experiences within the international network of biosphere reserves, an innovative type of protected areas which originated within UNESCO's Man and the Biosphere (MAB) Programme in the early 1970s. Biosphere reserves represent both a concept and a tool, addressing the conciliation of conservation objectives with development. As of mid-1994, there are 323 biosphere reserves in 82 countries, representing a total surface area of 211 million hectares.

The situation of individual reserves is variable. In some cases, the denomination of biosphere reserve has been added to that of an existing national park, with little change in emphasis or management philosophy. In others a real attempt has been made to integrate the multiple functions and participatory approaches that are intrinsic to the biosphere reserve concept. Experience in several biosphere reserves has served to highlight key management issues at stake in tropical savanna regions, as well as recent attempts to resolve resource use conflicts in particular localities. To the extent that the 'process is the product', the involvement of local people in discussions and decisions on issues such as those flagged in Table 1, can be considered a critical ingredient in the integration of community development, scientific research and conservation.

5.2 Policy insights

Several attempts have been made in recent years to rethink the assumptions and tenets that underpin savanna land use and management. This 'new' thinking reflects the differences between so-called equilibrium and non-equilibrium environments and increasing interest in approaches to conflict resolution and consensus building, and suggests several key principles for management and policy in areas characterized by high levels of variability which call for local level, flexible responses to uncertain events. Insights can also be gleaned from more general assessments of ecosystem management and resource use not primarily focused on savannas.

6. ECOLOGICAL CONSTRAINTS AND POLICY OPTIONS IN SAVANNA LAND USE

One set of guidelines and prescriptions has been developed as a joint operation of specialists from the social and natural sciences (Young and Solbrig 1992, 1993). The rationale is that the search for new policies should be based on a better understanding of existing management systems and land capabilities, including proper assessment of existing cultural and ecological constraints to development and of all the costs and benefits of proposed changes. Within such a perspective, policy opportunities include: strengthening social technology at the local level; formally documenting and registering traditional rights; adopting administrative arrangements which are ecologically sensitive to the episodic, variable nature of savanna systems; devolving responsibilities to local communities; and

Table 1: Management issues in a sampling of biosphere reserves in tropical savanna regions, and recent actions to address resource conflicts and reconcile conservation with development.

Biosphere reserve/management issues	Approaches to resolving resource conflict
Amboseli (Kenya). Increasing human and cattle populations, exacerbated conflicts in resource use and sharpened competition of cattle and wildlife for water and grass and increasing hostility by Maasai pastoralists to wildlife populations.	Setting up of semi-autonomous parastatal organization (Kenya Wildlife Services) with mandate to run the parks in Kenya. Introduction of measures for sharing revenue from park's entrance fees with local people living adjacent to park. Quarter of total gate entry fees allocated to community services (schools, health centres, water and cattle dips). Round table discussions of park managers and Maasai community on different management issues (e.g. water supply, compensation for wildlife-caused damage to farms in buffer and transition zones).
Boucle du Baoul Existing reserve boundaries and zonation patterns at odds with recent human activities (expansion of several villages in conservation area, use of area by transhument herders, extensive burning of vegetation).	Reduction of core protected reserve area by 33.7%. Rezoning of core, buffer and transition areas aimed at improved management of biodiversity resources, following series of ecological and socio-economic surveys completed in 1993. Diverse research in buffer zone on management of village territories and on biodiversity dynamics. Reinforcing national institution responsible for reserve management.
Cerrado (Brazil). Quarter of Brazilian territory covered by cerrado vegetation type (150 million ha, practically continuous in 13 Brazilian states), with large-scale transformation by agriculture (more than half converted to monocultures) and extensive livestock raising. Lack of awareness on biochemical and medicinal potential of cerrado plants. Only about 7% of cerrado `intact'.	Establishment in 1993 of 226,000 hectare Cerrado Biosphere Reserve in Federal District of Brasilia. Proposed first stage in nationwide network of protected 'Campas cerrados' sites, enjoining collaboration of local community organizations, municipalities, federal and state authorities, research institutions (i.e. similar strategic plan to that adopted for protecting remaining remnants of Mata Atlantica forest).
Comoe (northeastern Côte d'Ivoire). Pressures on protected areas for agricultural use by people living around the national park.	Seeking ways to improve low input agriculture in areas contiguous with the national park, through understanding and manipulation of biological processes affecting soil fertility. Within UNDP-UNESCO project on savanna productivity, field trials at nearby Bouna research station on effects of organic residue management on yields of maize, sorghum, peanuts.
Mount Nimba (Guinea). Growing pressure from iron-ore mining project, influx of refugees and local people wanting to farm in the area.	Mission in May 1992 organized by World Heritage Centre to determine exact boundaries of World Heritage site and impact of mining projects and other threats. In 1993, new site perimeters determined by Guinean Government for core area (= World Heritage site) within larger biosphere reserve. Environmental measures incorporated into mining project. Multilateral and binational support to continue protective measures and develop small-scale agricultural projects for local communities.
Niokolo-koba (Senegal). Multiple recent impacts and issues affecting integrity of Niokolo-koba (at the interface of Sudano-Sahelian and Sub-Guinean zones): siting of a tar-marked road across the national park; poaching of wildlife; and strong human pressure for land and use of park for dry season grazing.	In conjunction with events in 1994 to mark 40th anniversary of creation of national park, mobilization of national and international support for Niokolo-koba, with specific actions targeted at youth, local populations, NGOs and other partners. Revised management objectives, with measures for boosting local economy.

emphasizing community needs for regional rather than sectoral development.

One seemingly obvious conclusion of this particular assessment on policies for improved savanna management is that people should be recognized as an integral part of all savanna ecosystems and that the world's savannas should be studied from that viewpoint. In the past, the human dimension has been excluded from many ecological studies of savanna management. Similarly, the ecological dimension has been either excluded from most social and economic studies of the savannas or, alternatively, treated in a very simplistic manner.

A concluding set of general recommendations (Young and Solbrig 1992, page 42) suggest a need to: (a) begin with a careful assessment of the ecological capacity of savanna ecosystems; (b) avoid subsidies; (c) devolve responsibility for management to local communities; and (d) provide these communities with opportunities to make a significant but sustainable contribution to their economy.

6.1 Shifting priorities in African range management

Another set of policy directions have been generated in the context of African range management policies as a result of workshops held in Matopos, Zimbabwe, in January 1992 (Behnke 1992) and in Woburn, UK, in June 1993 (Behnke 1994, Scoones 1994), under the auspices of the Commonwealth Secretariat, the International Institute for Environment and Development and the Overseas Development Institute. Recognition that natural environments exploited by African pastoralists are generally robust and resilient is among the perceptions that have served to highlight contrasts between the 'old' and the 'new' thinking about pastoral development in Africa, and to suggest a series of principles for project and programme design in uncertain environments (Scoones 1994): recognize that long-time frames (15 years or more) may be needed for successful iterative planning and intervention with the involvement of pastoralists; start small and build up, focusing on institutional capacity at a local level; resist unrealistic disbursement targets; accept that projects are learning experiments, and thus be open to changing course if necessary and discarding outdated or irrelevant project plans; learn from experience (especially occasional episodic events) and gear monitoring and evaluation mechanisms to the rhythm of learning in variable environments; promote measures leading to institutional and organizational flexibility (required to allow responses to unexpected events) and dismantle bureaucratic project structures and procedures that stifle innovation; avoid being shackled with one organizational model and recognize that a diversity of different organizations may be appropriate to tackle complex challenges found in pastoral areas (pastoral organizations, service NGOs, producers' federations, government, ... all may have roles); and be aware that local level development is affected by macro-level policy, and acknowledge that tackling these wider issues (e.g. through support to legal cases, policy advocacy and lobbying) may be directly relevant to local level pastoral development.

7. POLICY CHALLENGES IN COMMERCIAL RANGELAND AREAS

Policy needs and challenges in commercial rangeland areas have also received a fair amount of attention from managers and researchers in such countries as Australia. Stafford Smith and Foran (1993), in stressing that these areas should not be seen as failed croplands, have argued that visions and goals need to be better articulated at both policy and industry levels, to 'envision the future' that is sought. The policy level must clarify the balance that society expects to be maintained between production and conservation in different regions; the regional industry level must identify the special values of production from different regions, and give rangelands production a separate identity to other agricultural production. These visions need to recognize the world trends towards ongoing net reductions in biological and economic returns, increasing environmental standards in a variety of ways, and greater interest in alternative rangeland uses, especially for conservation and recreation. The authors see an obvious niche for the commercial rangelands industry in relation to the remainder of agriculture: free-range, low residue meat and fibre produced from lands which do not compete for more intensive agriculture and which operate in harmony with other values and uses.

To meet this vision there are some important requirements, more particularly that production must be sustainable: in simple terms, this is achievable in an unreliable climate by a range of strategies from one extreme of matching stocking rates to forage supplies annually (resulting in fluctuating animal numbers) to another of operating at very low stocking rates which can cope with the driest of years. However, strategies tending towards the latter end of the spectrum are likely to be more compatible with minimizing errors, maintaining maximal marketing flexibility, minimizing reliance on uncontrollable externalities such as interest rates and trade conditions, and maintaining a constant, reliable level of production (Pickup and Stafford Smith 1993, Stafford Smith and Pickup 1993).

Notwithstanding this approach, there will still be environmental changes on the rangelands. Some of these are a result of the need to repair damage from the past, others will be deliberate or accidental manipulations in the future. The management of these changes will require industry to continue to innovate, using the principles of adaptive management (Holling 1978, Walters 1986). This in turn demands an increased investment in human capital and knowledge in the rangelands, as well as in modern technologies and approaches such as decision-support systems (Stuth and Lyons 1993) and in methods for motivating innovation which do not include subsidies to maintain past practices.

Fundamental pathologies of ecosystem management

Another set of policy insights have emerged from Holling's (1986, 1994) identification of a fundamental pathology in examples of ecosystem management, which suggests that any attempt to manage ecological variables (e.g. fish, trees, water, cattle) leads to less resilient ecosystems, more rigid management institutions and more dependent societies. Holling's (1986) review of twenty-three examples of managed ecosystems revealed a common feature and goal — that of controlling the variability of a target (e.g. insects and fire at low levels, cattle grazing at intermediate stocking densities, salmon at high populations) whose normal fluctuations imposed problems and periodic crises for pulp mill employment, recreation, farming incomes or fishing catches. In each case, the focus was on a narrowly defined problem. Modern engineering, technological, economic and administrative experience can deal well with such narrowly defined problems, and in each example the goal was successively achieved (e.g. cattle grazing managed with modern rangeland practice). At the same time, elements of the system were slowly changed as a consequence of the initial success of the policy. Because the problem was defined narrowly, such slow changes in the spatial heterogeneity of the ecosystem were not perceived (e.g. in terms of increasing contiguity of forest architecture over landscape scales, or the shift in rangeland composition between drought-sensitive and drought-resistant grasses). Disturbances that could previously be absorbed tended to flip the system into a persistently degraded state.

These changes in ecosystems — the trend to more spatially homogenized ecosystems over landscape scales, the loss of resilience — could be managed if it were not for concomitant changes in two other elements of the inter-relationships — in management institution(s) and in people in society who reaped the benefits or endured the costs. Because of the initial success, in each case the management agencies shifted their objectives from the original social and ecological ones to the laudable objective of improving operational efficiency of the agency itself — for example, in terms of the cost-efficiency of spraying insects, fighting fires, producing beef and releasing hatchery fish. Efforts to monitor the ecosystem for surprises rather than only for product therefore withered in competition with internal organizational needs, and research funds were shifted to more operational purposes. Thus the gradual reduction of resilience of the ecosystems was unseen by any but maverick and suspect academics whose research was driven simply by curiosity.

Success brought changes in the society, as well (Holling 1986). Dependencies developed and powerful political pressures were exerted for continuing sustained flow of the food or fibre that no longer fluctuated as it once had. More investments therefore logically flowed to expanding pulp mills, recreational facilities, cattle ranches and fishing technology. That is the development side of the equation, and its expansion can be rightly applauded. Improving efficiency of agencies should also be applauded. But if, at the same time, the ecosystem from which resources are garnered is becoming less and less resilient, more and more sensitive to large scale transformation, then the efficient but myopic agency and the productive but dependent industry simply becomes part of the source of crisis and decision gridlock. In sum, the very success in managing a target variable for sustained production of food or fibre apparently leads inevitably to an ultimate pathology of less resilient and more vulnerable ecosystems, more rigid and unresponsive management agencies and more dependent societies.

Within such a perspective, Holling has suggested that the pathology is broken when the issue is seen as a strategic one of adaptive policy management, which requires:

— integrated, flexible policies (not piecemeal, rigid, locked-in ones);

— management and planning for learning (not only for economic or social product);

— monitoring designed as a part of active interventions to achieve understanding and to identify remedial response (not monitoring for monitoring's sake);

— investments in eclectic science and in science at appropriate scales (not only in controlled science);

— citizen involvement and partnership to build 'civic science' (not public information programmes to passively inform).

8. Conclusion

The diversity of tropical savannas (in terms of climatic regimes, physical conditions, cultural and historical roots, and economic conditions) means that no single set of policy goals or management prescriptions can be offered for the 'proper use' of savannas. The level of human investment, the objectives of a given society, and the characteristics and previous history of a savanna will determine which policy frameworks and management regimes satisfy human aspirations and which do not.

While there may be no single vision for tropical savannas, there may well be widescale advantage in promoting new partnerships and connections between those whose activities affect land use and resource management in tropical savannas, including herder and farmer, research scientist and rangeland manager, economist and development planner. The diversities in savanna systems and actors are in turn nested in the broad array of the world's political, cultural and moral positions found in different parts of the world. Within such an array, a consensus that is affirmed by opposing theoretical, religious, philosophical and moral doctrines is likely to be more just and much more resilient than one based upon a single paradigm. The search is on for an overlapping consensus about the types of social and economic policies that promote sustainable forms of investment and resource use (Young 1992).

In terms of drawing lessons from the past for gaining insights to the future, the conclusions of Gunderson *et al.* (1994) bear repeating, rooted as they are in the dynamics of complex adaptive systems. Among the key lessons that emerged from their review of regional examples of coupled human and natural systems are the following: (a) the importance of recognizing and understanding the key interactions, across time and space scales, that provide robust solutions to crises and pathways out of 'gridlock'; (b) the significance of different management roles in enabling institutions to be more adaptive or flexible, in order to continually learn, to generate system understanding and to resolve the challenge of both simultaneously implementing and revising policy; (c) the need for more strategic, long-term, broad-scale and integrative policies, that serve to resolve a pathologic past and create credible futures; (d) the recognition that involvement and education of people who are part of the system are crucial to building resilient solutions and removing gridlock. Lessons such as these provide insights as to how to break-down barriers and to build new bridges for a sustainable future of complex adaptive systems of humans and nature, such as the tropical savannas.

References

Behnke, R.H. (1992). *New Directions in African Range Management Policy*. The results of a workshop held at Matopos, Zimbabwe, 13–17 January 1992. Commonwealth Secretariat, London.

Behnke, R.H. (1994). *Natural Resource Management in Pastoral Africa*. Commonwealth Secretariat-Overseas Development Institute-International Institute for Environment and Development, London.

Belsky, A.J. and Canham, C.D. (1994). Forest gaps and isolated savanna trees. An application of patch dynamics in two ecosystems. *BioScience*, **44**: 77–84.

Black, A.W. and Reeve, I. (1993). Participation in landscape groups: the relative importance of attitudinal and situational factors. *Journal of Environmental Management*, **39**: 51–71.

Casenave, A. and Valentin, C. (1989). *Les états de surface de la zone sahélienne. Influence sur l'infiltration*. Collections Didactiques. ORSTOM, Paris.

Casenave, A. and Valentin, C. (1992). A runoff capability classification system based on surface features criteria in semi-arid areas of West Africa. *Journal of Hydrology*, **130**: 231–249.

Chambers, R. (1983). *Rural Development: Putting the Last First*. Longman, Harlow.

Chambers, R. (1990). *Microenvironments Unobserved*. IIED Gatekeeper Series 22. International Institute for Environment and Development, London.

del Bono, E. (1988). *I Am Right, You Are Wrong*. Penguin Books, London.

di Castri, F. (1981). Ecology - the genesis of a science of man and nature. *UNESCO Courier*, (April): 6–11.

di Castri, F. (1992). Interview. *UNESCO Courier*, (September): 29.

di Castri, F. and Hadley, M. (1986). Enhancing the credibility of ecology: is interdisciplinary research for land use planning useful? *GeoJournal*, **13**: 299–325.

di Castri, F. and Hadley, M. (1988). Enhancing the credibility of ecology: interacting along and across hierarchical scales. *GeoJournal*, **17**: 5–35.

Gunderson, L.H. and Holling, C.S. and Light, S.S. (1994). Barriers broken and bridges built — a synthesis. In: *Barriers and Bridges in Renewing Ecosystems and Institutions*, (eds Gunderson, L.H., Holling, C.S., and Light, S.S.). Columbia University Press, New York (in press).

Hadley, M. (1993). Grasslands for sustainable ecosystems. In: *Grasslands for Our World*, (ed Baker, M.J.), pp. 12–18. SIR Publishing, Wellington.

Holling, C.S. (ed.). (1978). *Adaptive Environmental Assessment and Management*. International Series on Applied Systems Analysis 3. John Wiley & Sons, Chichester.

Holling, C.S. (1986). The resilience of terrestrial ecosystems, local surprise and global change. In: *Sustainable Development of the Biosphere*, (eds Clark, W.C. and Munn, R.E.), pp. 292–317. Cambridge University Press, Cambridge.

Holling, C.S. (1994). What barriers? What bridges? In: *Barriers and Bridges in Renewing Ecosystems and Institutions*, (eds Gunderson, L.H., Holling, C.S., and Light, S.S.). Columbia University Press, New York (in press).

Homer-Dixon, T.F. (1991). On the threshold: environmental changes as causes of acute conflict. *International Security* **16**: 76–116.

Kaplan, R.D. (1994). The coming anarchy. *The Atlantic Monthly* (February): 44–74.

Lane, C. and Scoones, I. (1993). Barabang natural resource management. In: *The World's Savannas: Economic Driving Forces, Ecological Constraints and Policy Options for Sustainable Land Use*, (eds Young, M.D. and Solbrig, O.T.), pp. 49-66. Man and the Biosphere Series 12. UNESCO and Parthenon Publishing, Paris and Carnforth.

Murphree, M.W. and Cumming, D.H.M. (1993). Savanna land use: policy and practice in Zimbabwe. In: *The World's Savannas: Economic Driving Forces, Ecological Constraints and Policy Options for Sustainable Land Use*, (eds Young, M.D. and Solbrig, O.T.), pp. 139-178. Man and the Biosphere Series 12. UNESCO and Parthenon Publishing, Paris and Carnforth.

Muul, I. (1993). *Tropical Forests, Integrated Conservation Strategies and the Concept of Critical Mass*. MAB Digest 15. UNESCO, Paris.

Pearce, D.W. (1993). Developing Botswana's savannas. In: Young, M.D., Solbrig, O.T. (Eds), *The world's savannas: economic driving forces, ecological constraints and policy options for sustainable land use*, pp. 205-220. Man and the Biosphere Series 12. UNESCO and Parthenon Publishing, Paris and Carnforth.

Pickup, G. and Stafford Smith, D.M. (1993). Problems, prospects and procedures for assessing the sustainability of pastoral land management in arid Australia. *Journal of Biogeography*, **20**: 471–487.

Renard, Y. (1991). Institutional challenges for community-based management in the Caribbean. *Nature & Resources*, **27**: 4–9.

Rittel, H.W.J. and Webber, M.M. (1973). Dilemmas in a general theory of planning. *Policy Sciences*, **4**: 155–169.

Sagasti, F. (1989). The new fractured global order. *Impact of Science on Society*, **155**: 207–211.

Scoones, I. (1994). *Living with Uncertainty: New Directions for Pastoral Development in Africa*. Overview paper of the workshop on New Directions in African Range Management and Policy. Woburn (UK), June 1993. International Institute for Environment and Development, London.

Stafford Smith, D.M. and Foran, B.D. (1993). Problems and opportunities for commercial animal production in the arid and semi-arid rangelands. *Proceedings of the XVIIth International Grassland Congress 1993*: 41–48.

Stafford Smith, D.M. and Pickup, G. 1993. Out of Africa, looking in: understanding vegetation change and its implications for management in Australian rangelands. In: *Rethinking Range Ecology: Implications for Rangelands Management in Africa*, (eds Behnke, R.H. and Scoones, I.), pp. 196-226. Overseas Development Institute and International Institute for Environment and Development, London.

Stuth, J.W. and Lyons, B.G. (eds). (1993). *Decision-Support Systems for the Management of Grazing Lands: Emerging Issues*. Man and the Biosphere Series 11. UNESCO and Parthenon Publishing, Paris and Carnforth.

Swift, M.J. and Bohren, L. and Carter, S.E. and Izac, A.M. and Woomer, P.L. (1994). Biological management of tropical soils: integrating process research and farm practice. In: *The Biological Management of Tropical Soil Fertility*, (eds Behnke, R.H. and Scoones, I.), pp. 209–227. John Wiley & Sons for TSBF and Sayce Publishing, Chichester.

Tabor, J.A. and Hutchinson, C.F. (1994). Using indigenous knowledge, remote sensing and GIS for sustainable development. *Indigenous Knowledge & Development Monitor*, **2**: 2–6.

Vaitilingam, R. (ed.). (1993). *Industrial Initiatives for Environmental Conservation*. Financial Times-Pitman Publishing, London.

Valentin, C. (1994). Soil erosion under global change. Paper presented at IGBP-GCTE First Science Conference. Woods Hole, 23–27 May 1994.

Walters, L.J. (1986). *Adaptive Management of Renewable Resources*. McGraw Hill, New York.

Webb, L.J. 1990. Beyond the forest. In *Australian Tropical Rainforests. Science-Values-Meaning*, (eds Webb, L.J. and Kikkawa, J.), pp. 117–123. CSIRO, Melbourne.

Young, M.D. (1992). *Sustainable Investment and Resource Use: Equity, Environmental Integrity and Economic Efficiency*. Man and the Biosphere Series 9. UNESCO and Parthenon Publishing, Paris and Carnforth.

Young, M.D. and Solbrig, O.T. (1992). *Savanna Management for Ecological Sustainability, Economic Profit and Social Equity*. MAB Digest 13. UNESCO, Paris.

Young, M.D. and Solbrig, O.T. (eds). (1993). *The World's Savannas: Economic Driving Forces, Ecological Constraints and Policy Options for Sustainable Land Use*. Man and the Biosphere Series 12. UNESCO and Parthenon Publishing, Paris and Carnforth.

Chapter 14

What Lies Ahead For The Tropical Savanna? Industries And Management Regimes

Mr Bill Gray AM,

Chief Executive Officer,
Office of Northern Development
Box 4075 GPO Darwin, NT 0801
Australia

1. Introduction

Over the next 30–50 years, many factors are likely to influence the development of Australia's savanna region, including: global changes to the world economy (e.g. the growing strength of Asia); changing international and domestic demand for goods and services (e.g. tourism); and increasingly tighter environmental policies. Existing and future industries will not be immune to these impacts and over the next 30–50 years, the nature and operation of savanna-based industries, including the mining, pastoral/agriculture and tourism industries, will change significantly. Given the multi-land use nature of the savannas, appropriate management regimes will be required to ensure the economic and ecological sustainability of the savannas and its industries.

1.1 Existing Industries

The savannas account for 65% of the land area of Northern Australia (the area north of 26°S) and its value may be gauged by considering the overall economic importance of Northern Australia. Despite its small population, Northern Australia generates a higher GDP (Gross Domestic Product) and more exports per capita than the Australian average. On the basis of 1989–90 statistics, when Northern Australia contained 5.4% of Australia's population, it accounted for 6.2% of GDP and 28% of exports (Centre for Applied Economic Analysis and Research 1992). Northern Australia is developing at a faster rate than the rest of Australia, with a heavy emphasis on tourism and resource industries (NT Department of Transport and Works 1994). Furthermore, the Bureau of Industry Economics (1994) reports that over the ten year period from 1981–82 to 1991–92, the Northern Territory recorded the greatest growth in GDP (57%) of all the States and Territories.

The savanna-based industries of mining, pastoral/agriculture and tourism currently represent the mainstay of the Northern Australian economy though their relative importance varies across the region (Table 1). In absolute terms, North Queensland dominates the contribution of these industries to the Gross Regional Output of Northern Australia (ASTEC 1993).

Mining is important to all three regions but more so for north-western Australia. Agriculture, on the other hand, is important primarily in north Queensland while tourism is becoming increasingly important for the Northern Territory. In addition, a considerable proportion of Northern Australia is under some form of Aboriginal control and other important land uses include national parks and areas reserved for Australian Defence Force activities.

The long-term sustainability and growth of these industries is essential for the continuing economic development of Northern Australia. To achieve this goal, future development strategies will need to consider the likely long-term business environment that these savanna-based industries will be expected to operate in.

Table 1: Gross Regional Product ($m) of the Regions of Northern Australia for the three Mainstay Savanna Industries[1], 1989–90

Industry	North Queensland	Northern Territory	North Western Australia
Mining	2 690 (51.0%)	847 (16.1%)	1 736 (32.9%)
Pastoral/Agriculture	1 211 (92.8%)	50 (3.8%)	44 (3.4%)
Tourism[2]	650 (69.2%)	230 (24.5%)	58 (6.2%)

1. The percentages in parentheses refer to the allocation of each industry across Northern Australia
2. Recreation, accommodation and personal services

Source: Braithwaite (1994)

1.2 Future Influences on the Savannas and its Industries

As mentioned previously, over the next 30–50 years, global changes are likely to influence the development of Australia's savanna region. Some of these changes were identified in a presentation on *Issues In Agricultural Research For Northern Development Over The Next 50 Years* given to the CSIRO Agricultural Advisory Committee in February this year and by a number of speakers at the recent ABARE Outlook '94 Trade Conference.

Significant social, population and political changes were forecast including an increasing cultural and economic integration with the rapidly growing Asian region. For example, Dengate (1994) predicts that in 30 years, 30% of Australians will carry some Asian genes and suggests there will be a significant amalgamation of Western and Asian cultures.

During the 1980s, the East Asia region experienced very high rates of growth. This growth is likely to continue throughout the 1990s and into the next century, although some easing of growth rates is likely as these economies mature (Fischer 1994). Consequently, as living standards rise in Asia, the demand for Australian mining and energy resources and agricultural commodities such as beef, sugar, dairy and horticultural products is likely to increase.

Australian business has already commenced the process of integrating with Asia. A recent survey of 'emerging exporters' found that over the last 20 years, the number of Australian manufacturing firms that chose East Asia as their first destination for exports had risen from 24%–51% (Australian Manufacturing Council 1993). In the future, this integration is likely to encompass Australia's resource, primary industries and services sectors.

2. How Will Industries Change In The Savanna?

The following crystal ball scenarios for some savanna industries are based on discussions held with a wide cross section of people with an interest in their development including representatives from industry, academia and government. Given that these scenarios take a 30–50 year outlook, they cannot be definitive and many people may disagree with their conclusions. However, they are provided as a basis for stimulating debate on the future of the savanna and its industries.

It is worth noting that Professor Harris of James Cook University concludes that it is very difficult to see the economy of Northern Australia changing its structure and course dramatically over the next couple of decades (ASTEC 1993). The population is too small, the area too large, minerals and agriculture too concentrated in specific locations, and manufacturing too dependent on mining and agriculture, to expect great change. While Professor Harris may be correct that the overall structure of the economy may not change, there will be major changes within the existing industries and growth in some emerging sectors. These changes will impact significantly on the use and management of the savanna.

2.1 Pastoral Industry

Expectations of a rapidly increasing demand for beef next century from a well developed Asia appear to have some validity. For example, the 17 million wealthiest Indonesians have a significantly higher per capita disposable income (in real terms) than the 17 million people who live in Australia (DPIF 1994), and increasing disposable incomes in South East Asia have improved its capacity to pay for relatively expensive animal protein such as beef. In addition, an analysis of potential markets (Table 2) indicates the demand for imported beef among both North East and South East Asian countries is likely to grow from the present 800 000 tonnes product weight to 1.6 million tonnes product weight by the year 2003 (Western Australian Department of Agriculture 1994).

Opinions differ on how the pastoral industry might change in the long-term to meet this demand. Some people in the beef industry have suggested that, if the Ord River area realised its full agricultural potential and became a major grain producing region, revolutionary changes would occur in the beef industry of Northern Australia. Under this scenario, the beef industry would almost entirely centre around feedlot operations and the large free range operations of today would gradually disappear.

One long term (10–20 year) value-adding proposal is to progressively replace live exports to South East Asia with chilled beef offered by a Northern Territory export works at prices competitive with live steer exports. An intermediate stage could be live cattle (giving greater value adding) replacing a large proportion of the live-feeder steer market (DPIF 1994).

Others in the beef industry argue that Northern Australia's comparative advantage depends on the provision of low cost grasses and that large pastoral operations will continue to be the most effective way of realising this advantage. In addition, the economic viability of establishing feedlot operations, capable of dealing with the climatic extremes of a region like the Ord, is questionable.

More than likely the savannas' future pastoral industry will assume a mix of both of these scenarios. Hence, whilst the most economic of the large pastoral stations will continue to supply the low cost beef and live cattle markets, future scientific and technological developments may assist the development of more value-added operations to supply the premium beef markets.

Dengate (1994) also canvasses some potential developments for the industry including: corporate owners adopting a nomadic-type system whereby cattle would be moved quickly throughout Northern Australia, as rainfall determines, to small tightly-managed areas of rich forage. The potential for this type of system may be increased if research breakthroughs can make local monsoonal grasses more suitable for cattle.

Land tenure is another issue facing the pastoral industry. Consumer concerns about the sustainable use of arid lands and recent changes to native title legisla-

Table 2: Potential Demand For Beef In North-East and South-East Asia ('000 tonnes product weight)

Country	1993	1998	2003
Japan	444.0	553.0	726.0
South Korea	143.5	238	337.3
Taiwan	49.9	74.5	99.0
NE ASIA	637.4	865.5	1 162.3
Hong Kong	60.5	91.3	114.5
Malaysia	50.4	86.0	132.8
Indonesia	19.0	62.0	98.0
Singapore	14.0	26.0	42.0
Thailand	0	23.0	61.0
Phlippines	11.0	N/A	N/A
Brunei	0.8	N/A	N/A
SE ASIA	155.7	288.3	448.3
TOTAL	793.1	1 153.8	1 610.6

Source: Western Australian Department of Agriculture (1994)

Table 3: Total Area of Aboriginal Land In Northern Australia[1], June 1994 (km[2])

	Aboriginal Reserves	Aboriginal Freehold	Aboriginal Leasehold	Multi-Manager[2] Use	TOTALS
Western Australia	143 564	14	101 040	nil	244 618
Northern Territory	nil	518 535	19 316	10 765	548 616
Queensland	2 957	21 141	18 977	nil	43 075
TOTALS	146 521	539 690	139 333	10 765	836 309

1. Northern Australia is defined as the land area north of the 26th Parallel of Latitude
2. Multi-Manager Use refers to the lease back arrangements developed for Kakadu National Park

Source: Data supplied by The Australian Surveying & Land Information Group

tion may eventually reduce the availability of land for pastoral activities. Currently, some 836 300 sq. kms or 23.8% of Northern Australia is under some form of Aboriginal tenure though this proportion varies across the region (Table 3). In the Northern Territory, for example, some 40% of the land is under some form of Aboriginal tenure.

Given current trends and the recent establishment of the National Aboriginal and Torres Strait Islander Land Fund, an increasing amount of the savanna will become subject to Aboriginal tenure with a subsequent increased involvement of Aboriginal interests in the pastoral industry.

There is already clear evidence of Aboriginal participation in the pastoral industry. Aboriginal pastoralists in the East and West Kimberley, for example, jointly commissioned a study of their future management, infrastructure and marketing requirements. As a result, they have endorsed a regional development strategy that involves a partnership approach with relevant Commonwealth agencies over an initial five-year period.

These actions confirm the growing recognition that Aboriginal interests are in for the long haul and that the forging of political and industry links between Aboriginal and other established pastoral interests will be a key ingredient for the future capitalisation and sustainability of the pastoral industry.

2.2 Mining Industry

The sustainable growth of the mining industry is vital for the future economic development of Northern Australia. Apart from the obvious economic returns to Australia arising from the continued export of raw mineral products, significant longer-term benefits could be achieved through value-added processing and the flow-on development of new industries in the savanna region.

However, to ensure this sustainability, the mining industry must come to terms with a number of issues, including: increasing community concerns about the environmental impact of mining; the need to conserve energy; and Aboriginal ownership of land and the implications of the recent native title legislation. In addition, the mining industry must work towards co-existence with other savanna industries and, through value-added processing, play an active role in development of new industries throughout the region.

Recent initiatives suggest that the mining industry has made considerable progress towards tackling these issues. For example, a number of mining and resource companies (e.g. CRA Pty Ltd and ERA Pty Ltd) have made substantial investments in the development of in-house environmental protection and remediation capabilities. In some cases, the success of these internal capabilities has led to the establishment of a separate commercial business that provides environmental services to external clients located both in Australia and overseas (e.g. ERA Environmental Services Pty Ltd).

Furthermore, some twelve months ago, the mining industry established an Australian Centre for Minesite Rehabilitation (ACMR) which consists of a network of research centres, across Australia, with special expertise in the field of minesite rehabilitation. The objectives of ACMR are to: address the major long-term strategic areas of minesite rehabilitation through coordinated research programs; enhance the training of postgraduate students in major research on minesite rehabilitation; and conduct regular training and extension courses for environmental and other professionals associated with the mining industry, dealing with best practices in minesite rehabilitation. Currently, the mining industry is seeking to strengthen the ACMR's capabilities through the fourth round of Commonwealth government's Co-operative Research Centre program.

A number of mining companies are also supporting research into alternative energy technologies with a view to creating low-energy mining operations that have negligible impact on the environment. Over the next 50 years, it would not be unreasonable to expect that solar and other alternative energy sources may play a large role in future mining operations.

The recent opening of the Mt. Todd Gold Mine in the Northern Territory, which involves Aboriginal interests, is another example of a potential trend for future

mining developments in the savanna region. Over the next 30–50 years, there is likely to be a significant increase in Aboriginal ownership in mining operations as the Aboriginal community strives for economic independence.

In a recent discussion paper, Altman (1994) stated that *'Aboriginal interests appear to be increasingly recognising that land alone will not result in regional economic development: capital, and human capital is also needed'*. Altman goes on to claim: *'A new form of joint venture with indigenous equity participation, employment and training opportunities and buy-back options is evolving. Ultimately it appears that when both parties are willing to move from confrontation to consultation to negotiation, positive outcomes will eventuate. Importantly, the absence of automatic royalty-equivalent payments under both agreements and the* Native Title Act 1993 *framework appears to be encouraging a more active Aboriginal involvement in resource development projects. There is also a growing Aboriginal recognition that the payment of royalty equivalents to incorporated bodies in areas affected by mining will often result in excessive regional politicking for these mining moneys, with a concomitant lack of attention to longer-term economic opportunities and an inability to accumulate venture capital for investment'*.

As Australia progresses into the next century, greater emphasis will be placed on value-adding processing and regional mining communities, like Mt. Isa, may have the opportunity to diversify their economic base. Some of these opportunities were highlighted in a recent case study of the Carpentaria and Mt. Isa Mineral Province (PA Consulting Group 1993).

The Carpentaria and Mt. Isa Minerals Province encompasses an extensive area of North West Queensland, centred around Mt. Isa, and the Northern Territory including the McArthur River mineral deposit near Borroloola. According to the study, known mineral deposits in the Province could, in aggregate, represent an additional AUD$25–30 billion in export revenue over the next 20–30 years and some AUD$2.0–2.5 billion in investment over the next ten years.

The study confirmed that remoteness and lack of established infrastructure, characteristics commonly associated with the entire savanna region, were impediments to development in the Province. Initiatives proposed to address these impediments, particularly in regards to energy and transport needs, include: the provision of cheaper energy, as opposed to that currently provided by coal and distillate, through the construction of a gas pipeline from South West Queensland to Mt. Isa; zinc concentrate being transported north in slurry pipelines and barged offshore for loading onto ships in the Gulf; and a reduction in freight charges on the Mt. Isa - Townsville rail line.

Other value added industries that could develop in the region include copper and lead smelting, fertiliser production and a Mt. Isa based heavy engineering and construction services industry. The regional implications of these developments for the savannas would include: a significant reduction in rail freight charges; a substantial change in the volume and mix of products handled by the Port of Townsville; significantly increased activity in the Gulf; increased copper refining in Townsville and a broader economic base for the Province. However, this and other similar proposals, will need the backing and commitment of the mining industry if they are to succeed.

2.3 Tourism

The tourism industry is a relatively small but regionally important contributor to the overall economy of tropical Australia. For the savannas, the industry's excellent prospects for future growth are based on a number of comparative advantages, including: the natural beauty and remoteness of the savanna landscape; numerous nature reserves and world class national parks (e.g. Kakadu and the Bungle Bungles); and the strong sense of ownership and attachment that the savannas have for many indigenous Australians.

A flow on benefit of tourism is that it can often provide resources for environmental conservation and management. However, unless properly managed, tourism can also result in damage to, or even loss of, the resources on which it depends. The future of savanna based tourism will depend on how well we come to terms with this possibility.

Over the next 30–50 years, there is likely to be a greater integration of tourism with other industries. Already, there is a growing trend for pastoral properties to diversify into the tourist market with El Questro Station in the Kimberley region being a prime example. There is also likely to be a significant growth in the number of tourist ventures operated and owned by Aboriginal interests. Increased land ownership arising from the Mabo decision, cultural affiliations and Aboriginal sensitivity to maintenance and custodianship of traditional lands will help drive this trend. It is worth noting that the recent Royal Commission into Black Deaths in Custody identified tourism as a possible source of Aboriginal employment.

On a regional basis, there is likely to be a greater integration of tourist activities across the savannas. For example, one concept currently being discussed by the Western Australian and Northern Territory governments is the formation of an 'Ibis Highway'. This concept involves a network of small aircharter operators and tourist operators providing an integrated tourism package that links Kakadu National Park with the remote Kimberley region.

As with the rest of Australia, there is also growing interest in ecotourism in the savannas. The Commonwealth government's national ecotourism strategy aims

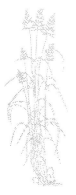

to provide a national framework for guiding the planning, development and management of tourism in natural environments while maintaining environmental quality (Commonwealth Department of Tourism 1994).

The government has allocated $10 million to implement this strategy and has recently announced funding for a number of specific projects in Northern Australia including: a study of the effects of boat tours on the presence and behaviour of wildlife in Kakadu national park in the Northern Territory; and the development of a regional ecotourism plan for the sustainable development of the Shark Bay world heritage area, Ningaloo marine park, and the Kennedy Ranges national park in the Gascoyne region of Western Australia

All of these actions highlight the growing importance that will be placed on the tourism industry to sustain the future development of the savanna.

2.4 Defence Activities

As mentioned previously, another important land use in the savannas involves areas reserved for Australian Defence Force activities. For both national and strategic reasons, Northern Australia will remain a primary focus for Australian defence planning and a focus for strategic engagement with regional defence forces in East Asia. The development of training facilities such as the proposed North Australia Training Area will have implications both in terms of sustainable land usage and the potential for greater joint exercise activities with regional defence forces.

2.5 Emerging Industries

A number of small but promising industries appear to be developing in tropical Australia especially in the fields of agriculture, horticulture and aquaculture. The long-term future for these industries will depend on the market demand for these products and the commitment of the people involved to meeting that demand.

The economic rationale behind these type of developments has been questioned for many years. For example, Davidson (1965) wrote: *'The Australian public should realise that agricultural development north of the tropic is expensive and would have to be supported by annual subsidies. Any of the crops which are produced there could be produced at lower cost south of the tropic. The only schemes which are economically sound are those aimed at improving the existing large-scale pastoral industry'*.

Despite this pessimism, positive developments are now occurring and with future scientific and technological advances, may well create a range of export oriented industries in the savanna region. This potential was highlighted in a recent study of the Ord River Irrigation Project (Hassall & Associates Pty Ltd 1993).

The study indicates that there has been a significant growth in the level and diversity of horticulture production in the Ord District in recent years. For example, during the five years from 1985–86 to 1990–91, the gross value of horticultural production from the Ord increased from approximately AUD$6–$25 million. In addition, the range of crops grown in the Ord has expanded from sorghum, sunflower and forage crops in the 1970s to include now maize, soybeans, peanuts, hybrid seed, chickpeas, mangoes and bananas. The study suggests that depending on the rate of expansion over the next three decades internal rates of return could be in the range of 30%–60%.

These dramatic improvements are attributed to: the development of specialised field crops and horticultural industries; developments in biological control of insect pests; the marketing skills of the Ord River District Co-operative; the tenacity of farmers and the growth in Asian markets. Furthermore, the Western Australian government has embarked on a survey of Stage 2 of the Ord River development, wherein the remaining two-thirds of irrigable land will be brought into production.

The growth of the Asian economies is also likely to impact on the agricultural profile of the savanna region. As ASEAN countries will start importing rice around 2000, having exhausted their land reserves on urban expansion and development, prices will rise despite dietary substitution. This, in turn, will probably generate considerable Asian-financed irrigated production in the expanded Burdekin and Ord-Keep River districts, integrated with beef finishing and freshwater aquaculture (Dengate 1994).

The success of these and other industries will be enhanced if they focus on products which are naturally suited to the tropics. One possibility is the development of an efficient export sugar industry in the Ord region.

CSR Pty Ltd have just announced that they will invest $30 million in the development of a raw sugar mill in the Ord. The mill will be a prototype which involves significant research and development in its construction and testing phases and is being supported through the Commonwealth government's 150% R&D Tax Concession Scheme. The new operation, comprising 24 of the Ord River District Co-operative's cane growers and CSR as the sugar miller, is planning to grow 560 000 tonnes of cane and produce 70 000 tonnes of sugar annually. Land is being planted immediately to produce up to 100 000 tonnes of cane for mill trialling and testing in August 1995 with the 1996 season expected to be the first year of full production.

In the short term, this development may establish an export sugar industry and associated employment opportunities in a key developing region of Australia. More importantly, in the long term, the project could help create a new and internationally competitive export industry based on the manufacture of sugar mills.

Table 4: Northern Territory Legislation Covering the Pastoral Industry	
Administering Authority	**Legislation**
Dept. of Lands and Housing	*Pastoral Land Act* *Crown Lands Act* *Special Purpose Lease Act*
Dept. of Primary Industries and Fisheries	*Abattoir and Slaughtering Act* *Biological control Act* *Brands Act* *Exotic Diseases (Animals) Compensation Act* *Meat Industry Act* *Noxious Weeds Act* *Pet Meat Act* *Plant Diseases Act* *Poisons and Dangerous Drugs Act* *Seeds Act* *Stock (Artificial Breeding) Act* *Stock (Control of HGP) Act* *Stock Diseases Act* *Stock Routes and Travelling Stock Act* *Veterinary Surgeons Act*
Northern Territory Conservation Commission	*Bushfire Act* *Environmental Assessment Act* *Heritage Conservation Act* *National Trust (NT) Act* *Soil Conservation and Land Utilisation Act* *Territory Parks and Wildlife Conservation Act*

The future operation of these emerging industries across the savannas is likely to involve an unprecedented high level of inter-industry cooperation and networking. For example, there would appear to be many opportunities for the aquaculture, tropical fruits, exotic poultry (e.g. magpie geese) and other emerging industries to co-operate in the future. Already, we have seen a number of successful export-oriented networks established in the food field including the Cairns Food Group. This trend will increase and may assist the growth of other regional centres in Northern Australia such as the Ord River and Douglas-Daly Districts.

3. APPROPRIATE MANAGEMENT REGIMES

The savannas are an area of strategic significance and economic potential. Balanced resource management is essential if we are to optimise the benefits of future development. As mentioned previously, many factors will inevitably influence the future use of the savannas and we may need to develop appropriate management regimes to ensure the savannas' economic and ecological sustainability. The capability to resolve conflicts between the various stakeholders utilising the savannas will be a key feature of any such management regime.

Whilst recognising that there are many permutations and combinations of management regimes that might be considered for the savannas, the focus of this paper is on the relative merits of voluntary options as opposed to regulatory options.

3.1 A Regulatory Option

There is already a substantial regulatory framework governing the use of the savannas and the industries operating in it. In preparing this paper, an attempt was made to document and quantify the current legislation and regulations covering the key industries and their use of the savannas. However, it quickly became evident that this supposedly simple task was going to require substantial resources to sort through the regulatory labyrinth that currently exists across all three levels of government.

By way of example, the local legislative framework covering just the pastoral industry in the Northern Territory involves at least three Territory government agencies in the administration of 24 Acts of Parliament (Table 4). This does not take account of Commonwealth and local government legislation that also governs the pastoral industry. For the savannas, separate but similar prolific levels legislation are to be found in Queensland and Western Australia. Furthermore, a similar and confusing level of legislation and administrative arrangements applies to other savanna industries such as mining and tourism. Needless to say, many people may question the efficiency of such a complex regulatory framework.

One option could be to rationalise and centralise this framework by establishing a single regulatory body to oversee the management of Australia's savanna region. This could be analogous to the establishment of the Great Barrier Reef Marine Park Authority (GBRMPA) or the Murray Darling Basin Commission (MDBC).

The concept of an overarching regulatory authority overseeing the management and development of the savanna region offers some advantages, including:

- improved inter-agency communication and avoidance of confusing policy differences between the three levels of government and across the States and Territory;

- resources being rationalised and focused on the most strategically significant issues; and
- improved co-ordination of efforts to attract the level of funding required to tackle large scale problems, like restoration, that require multi-disciplinary and multi-property solutions.

However, in establishing the GBRMPA and MDBC management models, there was a clear need to resolve cross-jurisdictional dilemmas where action or inaction in one jurisdiction had direct implications for another. For example, irrigation practices in Victoria and New South Wales had a direct impact on the water quality of the Murray River flowing into South Australia.

It is doubtful that such distinctive and contentious cross-jurisdictional issues exist in the savanna region and it may not be appropriate, let alone practicable, to manage the savannas as a single discrete region of Australia. The difficulties of this scenario are also compounded by the number of governments and diversity of industries and stakeholders involved. This does not mean that the operation of industries in the savannas would not benefit from the rationalisation of the regulatory framework.

3.2 A Voluntary Option

Another, more preferable option would be to establish a broad management consultative network across the savannas. This option is based on the premise that local communities and industries are capable of delivering sustainable management of the savanna region if they are provided with the necessary information and resources. This would be similar to the successful approach employed in the National Landcare Program (NLP).

By way of background, the NLP provides for an integrated approach to encourage responsible management of Australia's land, water and living natural resources. Commonwealth funding is available to community groups involved in identifying and solving their shared natural resource problems in a co-ordinated and ecologically sustainable sound manner.

The program arose from the joint effort of the National Farmers Federation and the Australian Conservation Foundation to tackle the issue of land degradation and is an excellent example of co-operation between industry and the conservation movement. Similar co-operation between the many land users must be cultivated if we are to realise the savannas' full potential.

In considering the sustainable development of the savannas, there appears to be many synergies with the current philosophies and aims of the Landcare Program. For example, in supporting the NLP, the Commonwealth believes that profitable and internationally competitive pastoral, mining, agricultural, tourism and other rural industries will continue to play a key part in Australia's economic future.

The Commonwealth's Landcare vision is to raise the long term productivity and ecological sustainability of our land resources and thereby ensure a future for rural industries. To achieve this balanced and sustainable use of land, the Commonwealth works towards:

- all land managers being able to make well informed decisions based on an understanding of the full economic, ecological and social costs and benefits of their land use practices;
- institutional arrangements which clearly put the responsibility for addressing land degradation with those who cause it;
- a research, development and demonstration effort that promotes sustainable, internationally competitive and efficient rural industries in the long term;
- all levels of government, the community and individual landholders understanding the nature and value of our land resources and working in partnership towards their sustainable use; and
- reconciling economic, social and ecological concerns in the management of Australia's land to best sustain a full range of uses for the benefit of the whole community into the future (Commonwealth of Australia 1991).

There appears to be considerable scope to employ similar strategies in the future management of the savannas. In fact, the establishment and development of local community-based groups to manage the savannas may best be achieved under the auspices of the current Landcare Program. These local groups would aim to achieve a long term commitment to sustainable development of the savannas by communities and the individuals in those communities. However, to be successful, this option will require the voluntary commitment of interested parties and maximum use should be made of existing networks already established in Northern Australia (e.g. the Landcare groups, Industry Associations and Aboriginal Land Councils).

This option is also consistent with current government thinking on regional development whereby regions are being encouraged to take a lead in shaping their own future. Also, like the Landcare model, the establishment of a management consultative network for the savannas may require catalytic funding and resources from the Commonwealth government.

4. INFORMATION AND R&D NEEDS

Regardless of which management option is pursued, reliable and accurate information on the ecological, economic and cultural impacts of multiple land use is

required if we are to make informed and sound decisions on the management of the savannas. Such information is not currently available and little is known about the sustainable capacity of this fragile and remote landscape with its harsh environment of natural extremes.

In November 1993, a report on *Research and Technology in Tropical Australia: Selected Issues* was submitted to the Prime Minister's Science and Engineering Council (PMSEC). The report concluded that:

'*Current research in the region is piecemeal and poorly integrated between the different client groups. Future research must address the identified needs of the savanna landscape as a whole. These needs must be determined by the regional community and promote intergovernmental cooperation to ensure commitment to long term sustainability*' (PMSEC 1993).

The Office of Northern Development (OND) supports this conclusion and believes it is valid for the whole of Northern Australia. Furthermore, we believe that the various stakeholders with an interest in Northern Australia (e.g. the mining, tourism and pastoral industries, Aboriginal groups, the Military and Environmental groups) need to be involved in the open determination of future research priorities.

OND also supports the need to improve the focus, coordination and communication of research in Northern Australia. More specifically:

a. Focus is required to ensure that research is directed at the economic, social and environmental needs of Northern Australia. This includes the needs of key stakeholders such as the Aboriginal community and existing and emerging industries. Currently, there is no comprehensive mechanism to ensure that research agencies are targeting these needs.

b. Co-ordination of research priorities between the various research agencies is needed to minimise duplication and gaps in the overall research effort and to maximise the value of the research dollar. This is particularly important in Northern Australia where small groups of researchers are trying to cover many important issues over a vast area.

By their very nature, the key industries of Northern Australia impact on each other. For example, the environmental effects of the mining and pastoral industries can impact on the viability of the tourism industry. Co-ordinated research will enable these flow on impacts to be taken into account.

Similarly, we need to ensure that social and economic research in Northern Australia is co-ordinated. This is particularly important for the many small and remote communities where social and economic development are highly integrated. Some progress has been made towards this goal as evidenced by the establishment of the North Australia Social Research Institute, which operates out of James Cook University in North Queensland, the Northern Territory University, and Hedland College in the North-West of Western Australia.

c. Communication needs to be improved to ensure that information on the research priorities most important to the development of Northern Australia is transferred to those undertaking research. Similarly, we must ensure that research results are disseminated quickly and in a user-friendly manner to all potential clients.

Broad dissemination is required to maximise the benefits of research results. For example, research on pasture improvement for the pastoral industry may also be useful to the mining industry for minesite rehabilitation and the tourism industry for minimising the environmental impacts of tourists. In the past, these other industries were not usually considered when the research results were being disseminated.

Current initiatives such as the proposed Co-operative Research Centre (CRC) for 'the Sustainable Development of Tropical Savannas' should be encouraged and, if successful, should play a key role in the development of a research network for the savannas. The proposed CRC aims to draw together a wide range of expertise from throughout Northern Australia, including key industry and public sector agencies (e.g. the Meat Research Corporation), to address strategic issues facing the savannas.

The objective of the proposed CRC is to provide management options and technology for the sustainable, multi-purpose use of tropical savannas by:

- quantifying and evaluating the effects of land uses on ecological processes across a range of spatial and temporal scales;
- developing technology to assess and predict the economic, ecological and social impact of land use and management at scales relevant to land users; and
- providing education and information, both user and public oriented, to empower people to make decisions that will lead to ecologically sustainable development.

Current efforts by the Land and Water R&D Corporation and the Australian Nature Conservation Agency to address some of these network issues must complement, rather than duplicate, any local initiatives such as the proposed CRC.

5. CONCLUSIONS

Over the next thirty to fifty years, we are going to witness considerable expansion and change in the nature and operation of the industries of the savannas

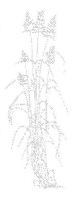

including the emergence of some promising new ones. To ensure this change is for the better, appropriate management regimes, that take into account the interests of all stakeholders, must be developed and they must evolve from the desires of local communities and industries rather than being imposed upon them.

If this is achieved, the future for the savannas should be one of optimism, growth and rational management generated and controlled by those who will be most affected.

References

Altman, J.C. (1994). *Implementing Native Title: Economic Lessons from the Northern Territory*. Discussion Paper No. 64/1994, Centre for Aboriginal Economic Policy Research, Australian National University, Canberra.

ASTEC (1993). *Research and Technology in Tropical Australia and their Applications to the Development of the Region*. Final Report, AGPS, Canberra.

Australian Manufacturing Council (1993). Emerging Exporters - Australia's High Value-Added Manufacturing Exporters. Final Report of the Study by McKinsey & Company and the Australian Manufacturing Council Secretariat, June.

Bureau of Industry Economics (1994). *State Economic Performance 1981-82 to 1991–92*. Occasional Paper 19, AGPS, Canberra.

Braithwaite, R. W. (1994). Working with a recalcitrant land: maintaining conservation value and improving economic production of Australia's northern lands. *Proceedings of the 8th Australian Rangeland Society Conference*, Katherine 1994, pp 89–96.

Centre for Applied Economic Research and Analysis, (1992). *Northern Australia Economic Development Strategy: Report to the Commonwealth Government*. James Cook University, Townsville.

Commonwealth Department of Tourism (1994). *National Ecotourism Strategy*. AGPS, Canberra.

Commonwealth of Australia (1991). *Decade of Landcare Plan - Commonwealth Component*. AGPS, Canberra.

Davidson, B. R. (1965). *The Northern myth: a study of the physical and economic limits to agricultural and pastoral development in tropical Australia*. MUP, Melbourne.

Dengate, H. (1994). Issues in Agricultural Research for Northern Development Over The Next 50 Years. Paper Presented To The CSIRO Agricultural Advisory Committee, 21 February.

NT Department of Primary Industry and Fisheries (1994). A Scenario for the Northern Territory Beef Industry for the Decade 1993-2003. Draft Paper, February.

Fischer, B. (1994). *Commodity Overview*. Paper Presented to the ABARE Outlook 94 Conference, Canberra 1-3 February, Conference Proceedings Vol 1.

Hassall & Associates Pty Ltd (1993). Ord River Irrigation Project, Past, Present and Future: An Economic Evaluation. Prepared for the Kimberley Water Resources Development Office, August.

NT Department of Transport and Works (1994). Draft Northern Territory Transport Strategy. Prepared by the Transport Division, March.

PA Consulting Group (1993). Carpentaria and Mt Isa Mineral Province Study: Conclusions, Recommendations and Actions Arising from the Study. Prepared for a joint Commonwealth, Queensland and Northern Territory government/industry group chaired by the Premier of Queensland, 26 March.

Prime Minister's Science and Engineering Council (1993). *Research and Technology in Tropical Australia: Selected Issues*. Office of the Chief Scientist, Department of the Prime Minister and Cabinet, AGPS, Canberra.

Western Australian Department of Agriculture (1994). The Kimberley Beef Project -Kimberley Beef Industry Business Plan — Part One. Kimberley Beef Industry Development Team, Miscellaneous Publication No. 7/94.

Chapter 15

A 2020 Vision for Cape York Peninsula; A Story of 40 000 years plus 200.

Matthew Baird

Cape York Land Council
Box 2496 PO Cairns, QLD 4870
Australia

Abstract

Cape York Peninsula has attracted international attention for its cultural and environmental values. One of the most significant planning exercises in the world, with publicly funded participation by community groups, is currently being undertaken in the region. The Cape York Land Use Strategy, or CYPLUS, is nearing the end of its first stage of operation, the collection and analysis of scientific, technical and community data. Stage II, the development of the planning options, will commence at the end of 1994. Stage III, the selection and implementation of these options, will commence after that.

The future of Cape York Peninsula and the fulfilment or disappointment of the expectations of the residents of Cape York will be decided during the next few years. It is appropriate in this context to examine the future of Cape York Peninsula and the many who need to be accommodated.

Following the recognition of native title in the High Court in Mabo and others v. The State of Queensland (1992) 175 CLR 1 and the passage of the Native Title Act 1993 (Cth) a new dimension has emerged with the repudiation of terra nullius that has enabled aboriginal people to exert a moral pressure on governments and the community to redress a number of the historical wrongs committed against aboriginal people. There now exists a legal right of recognition for native title, and the consequent recognition of traditional and customary rights and obligations of aboriginal people.

In Cape York the aspirations of aboriginal people are all encompassing and include environmental, pastoral, tourist and developmental interests. In light of the current developments an extraordinary opportunity for creative solutions and positive outcomes has arisen.

Only if there is a genuine recognition of aboriginal aspirations for self management and self governance will Cape York Peninsula survive, not only for benefit for indigenous Australians but for the late comers as well.

1. THE HISTORICAL PERSPECTIVE

It was 1603 that Captain Janz, a Dutch sea farer, became the first recorded European to sight Cape York Peninsula. This he did on the West coast of Cape York. Though he found nothing to recommend the place, his association with Cape York continues today with such places as the Coen River being named by him. However the modern history of Cape York Peninsula begins really with Captain James Cook and the beaching of the Endeavour on the banks of the river that now bears that name at Cooktown or as the Aboriginal camp was known then, Gangarr.

Cook recorded in his journal (Hawkesworth 1795) (referring to the Gugu Yimithirr inhabitants of the region):
'They live in tranquillity which is not disturbed by the inequality of condition. The earth and sea of their own accord furnishes them with all things necessary for life. They live in a warm and fine climate and enjoy every day so that they have very little need for clothing. In short, they seem to set no value upon anything we gave them; nor would they ever part with anything of their own. This in my opinion argues that they think themselves provided with all the necessities of life.'

Yet a century later the editor of Cook's Journal, Captain Wharton (1893), described Aboriginal people in another way:
'Their treachery, which is unsurpassed, is simply an outcome of their savage ideas, and in their eyes is a form of independence which resents any intrusion on their land, their wild animals, and their rights generally. In their untutored state they therefore consider that any method of getting rid of the invader is proper. Although treated by the coarser owner of colonists as wild beasts to be extirpated, those who have studied them have formed favourable opinions of their intelligence. The more savage side of their disposition being, however, so very apparent, it is not astonishing that, brought into contact with white settlers, who equally consider that they have a right to settle, the Aboriginals are rapidly disappearing.'

In presenting a paper on the future one is trying to pierce the veil that forever shrouds the maybe from the will be. To do so effectively one must note that we proceed to that future not in some vacuum but within an historic context. As Noel Pearson, the Executive Director of the Cape York Land Council, noted in a recent speech to a history conference in Melbourne: 'Only in the past two decades has this country's historiography begun to approach the presentation of the past with some willingness to challenge and deconstruct the prevailing conventional ideology. This ideology produced a rosy historical panorama that could only be seen through the contact history lens of the colonists. In this country this visitor saw an unpeopled continent, ripe for European civilisation. This visitor saw only that promise said to be unavoidable, as being inherent to and one of the pitfalls of history; that the weak should inevitably lose and the stronger should prevail absolutely.'

Yet times and ideas are changing. Again in the words of Noel Pearson: 'The works of the leading historian Professor Henry Reynolds, and others who have revolutionised antipodean historiography have now illuminated the other side of the frontier. There is now a movement toward a history that is deconstructing that colonial visitor that is not concerned with preserving and reinforcing conventional colonial ideology, and is instead reaching forward to a contemporary understanding, unrestrained by the distortions of past constructions.
These past constructions justified colonialism and its reality in the present. The new history is not so concerned with reinforcing that colonialism that still lives amongst us today. The fact that colonialism is still tenaciously the prevailing ideology in this country is plain to see.
The government and business still adhere to this ideology. The natives of the continent still rightly fear colonialism. The environment is still being colonised. Aboriginal people in remote Australia still face prospects that are distinctly colonial. The mining industry in this country is still colonial. Indeed the industry is egregiously colonial, it is Livingstonian in its outlook, it is Kiplingesque, in its one eyed obsession with empire. It is just that the pith helmets are now made of plastic.'

The issues of the new history, and the issues raised by the presenters at this symposium such as Darryl Pearce and Donna Craig, are confronting. They are meant to be.

In his 1993 Boyer Lecture, Noel Pearson referred to the 1968 Boyer Lecture of the eminent Australian anthropologist Professor Stanner:
'His lectures articulate, illuminate and provide some guidance with questions that will consume the people of this continent for as long as we need to consider them. Such questions as the future of nationhood and the equivocal citizenship of those with whom a settlement of great grievance remains outstanding; indigenous rights and notions of new partnerships; constitutional and institutional renovations; the never decreasing need for social reconstruction and renewal; and the great imperative for equitable distribution of the hitherto not so common wheel.'

So where does that lead us? Well, it leads us to today. And it is today that we look to the future.

2. BRIEF DESCRIPTION OF CAPE YORK PENINSULA

The Cape York Peninsula area covers 130 000^2 km, most of which can be classified as tropical savannas, and has a population of approximately 14 000 people, 8000 of which are Aboriginal or Torres Strait Islanders. There are over 20 different communities with the main centres of population being Weipa (population 2000) and Cooktown population 1700. In addition there are numerous settlements scattered throughout the Cape.

Approximately 54% of the Cape York Peninsula area is covered by Pastoral Lease, 20% by Aboriginal or

Islander land, 12% by national parks and the rest by freehold and vacant crown land.

2.1 Issues for non-indigenous Australians

There are two specific issues that need to be addressed in the context of a vision of Aboriginal people in Cape York Peninsula. The first is the need for non-Aboriginal Australians to recognise an intrinsic Aboriginal ethic for conservation. The second is the need for non-indigenous Australians to see as valid the desire and demand by Aboriginal people for genuine involvement in land management practices.

3. AN ABORIGINAL CONSERVATION ETHIC

The question of an Aboriginal conservation ethic is particularly difficult. In an area such as Cape York Peninsula where ecological values rank amongst the highest in the world it is even more difficult.

An area the size of Victoria, the Cape is arguably this country's most important remaining wild region. The natural and cultural diversity is significant.

The Cape contains:

- Seventeen major river systems flowing into the Gulf and the Great Barrier Reef, one of these is Australia's second largest catchment after the Murray Darling;
- Australia's largest remaining area of lowland rainforest at Iron Range;
- Some of our most extensive species diverse mangrove systems;
- The world's largest dugong breeding grounds,
- Queensland's most important crocodile breeding sites;
- The world's largest tropical savanna range lands,
- Wetlands that rival Kakadu;
- Australia's highest concentration of rare and threatened plants;
- A mosaic of ecosystems, the diversity of which is unmatched in the defined region elsewhere in Australia;
- Bird and mammal species found nowhere else in this country;
- Significant parabolic sand dune systems, some types of which are unique to the world;
- The most apparent evidence of our faunal and floristic links with Papua New Guinea.

At the time of European settlement, it is estimated that the Cape contained the highest population density of indigenous people on the continent. The spiritual, historical and traditional links with country are still very strong.

In addition the east coast of Cape York Peninsula directly abuts the northern section of the Great Barrier Reef Marine Park and the southern part of Cape York contains areas of the World Heritage listed Wet Tropics region.

In a submission by the New South Wales Aboriginal Land Council to the House of Representatives standing committee on Environment, Recreation and the Arts in its inquiry into biological diversity in October 1991 it is stated:

'The natural environment — the land and the plants and the animals which live on it — is the basis of traditional Aboriginal culture, spirituality and life itself. The modern concept of ecology, which recognises the relationship of organisms to each other and their surroundings, was something every Aboriginal person understood deeply.

The dignity of traditional Aboriginal life is based in the most ancient and intimate knowledge of, and relationship with, the Australian continent. Its plants, animals and geology, its weather and tides were all known in the greatest detail by our ancestors. The Aboriginal people were the supreme ecologists and the supreme landcarers.

In return the land was cared for. The vast arid interior, ancient soils, sparse vegetation, and unreliable climate together produced a fragile balance which our people recognised and respected. The nomadic way of life and small population did not overburden the ecosystem. Resources were never depleted to the point of exhaustion before our people moved on to allow the area to recover. Killing was restricted to food only and the killing of young animals and breeding females was prohibited.

The essence of religious belief was oneness with the land and all the creatures upon it. A view which the human and other natural species were all part of the same life force. The ultimate proof of this belief is the total devastation of Aboriginal culture which took place when land was dispossessed.

The dispossession of Aboriginal land directly paralleled the destruction of Aboriginal life, language and culture, so inseparable are they. As white settlement pushed inland, so too was the Aboriginal world shattered. Traditional culture survives today where the white graziers stopped.'

It is in this context that we can also consider the indigenous involvement in the UNCED Convention on Biological Diversity. Australia's National Strategy on Biological Diversity (Australia 1993) includes Objective 1.8, which is to:

'Recognise and ensure the continuity of the contribution of the ethno-biological knowledge of Australia's peoples to the conservation of Australia's biological diversity.'

The draft ATSIC Environmental Policy focuses on biological diversity and proposes a strategy to protect native title rights and interests and biological diversity. The strategy which recognises the protection of our diversity acknowledges that continued subsistence uses, cultural uses, commercial uses and management of resources need to be consistent with the sustainable use of our diversity while not impairing native title rights. Traditional agricultural knowledge and traditional law are recognised as valuable in efforts to conserve plant and animal species.

4. Aboriginal Involvement in Management

The second important issue to recognise is the desire for Aboriginal people to be more completely involved in land management issues.

In this context it would serve us well to pause to examine what is currently happening in Cape York Peninsula. In terms of land use planning the answer is one of the most significant planning exercises in the world with government funded participation by a number of community groups. This is known as CYPLUS — the Cape York Peninsula Land-Use Strategy.

4.1 Brief description of CYPLUS

CYPLUS is a three to five year planning and land use study. The Commonwealth and Queensland Governments have each contributed half of a nine million dollar budget. CYPLUS has three major programs. The first is a natural resource analysis programme, the second a land use programme and the third a public participation programme. The natural resources analysis programme has been under way for the last two years. Its primary purpose is to assess and analyse natural resources in Cape York. The information gathered will be entered onto a GIS and will be available to use for mapping and charting purposes. The Land Use programme consists of 26 projects that aim to establish further information base beyond natural resources. These 26 projects which are currently being conducted in Cape York are there to flesh out the values, expectations and needs of the residents of Cape York Peninsula.

The final leg of CYPLUS is the public participation programme. One million dollars has been allocated for public participation in CYPLUS.

The public participation programme is an extraordinary experiment in attempting to involve communities within the planning process. The main aim of the public participation process is to fund a number of community groups to provide input and advice both on project formulation and on methodology and results. Six groups are currently funded under CYPLUS public participation programme. These are:

- the Port Kennedy Association (representing the Prince of Wales islands);
- the Cape York Peninsula Department Association;
- the Cape York Peninsula Pastoral Advisory Group;
- the Cairns and Far North Environment Centre;
- the Cook Shire Council;
- the Cape York Land Council.

Each of these groups is contracted to represent a number of groups or a specific interest group. For example the Cape York Land Council is a lead agency designed to enable aboriginal participation and input into CYPLUS. The Land Council is therefore contracted to consult with the community and to make sure that community input comes back to CYPLUS. This does not preclude aboriginal people or aboriginal interests being represented but it does provide a mechanism by which a lead agency takes a role in providing issues, information, and views to the whole planning process.

4.2 CYPLUS — Land Use Planning Committee

One of the important outcomes of CYPLUS has been the development of a proposed land use committee. At the conclusion of Stage One CYPLUS it is proposed to establish a Stage Two Land Use Committee that will assess and examine options for land use planning and development in Cape York Peninsula. It is proposed that this committee be comprised of at least 50% by Aboriginal and Torres Strait Islander people. This is an important acknowledgment of the recognition by non-indigenous people that Aboriginal and Islander people desire specific involvement in policy making bodies that affect land use practice.

4.3 Aboriginal management

Additionally at a workshop held at Pajinka from the 26th–28th April 1993, co-ordinated by the Cape York Land Council and the Great Barrier Reef Marine Park Authority, traditional owners expressed significant desire to be involved in a practical way with the management of the Great Barrier Reef Marine Park.

The concept of joint management involving Aboriginal people and non-Aboriginal people is a difficult one. From a scientific point of view it is often argued that Aboriginal people do not possess sufficient technical expertise to effectively make management decisions. This is particularly so in the case of modern problems that are often not within the scope of traditional management practices.

Aboriginal people have rejected that view specifically.

Indeed it was acknowledged by almost all participants that the primary cause of many of these management problems are not traditional activities but that one way of dealing with the problems is to limit and impose barriers to the carrying out of traditional activities.

This is particularly the case in Turtle and Dugong management. One of the prime causes of species loss for turtle and dugong is commercial fishing. However, much of the attention that has been focused on managing turtle and dugong has been focused at limiting catches and traditional use of turtle and dugong by Aboriginal people.

At the Pajinka workshop it was made fairly clear that Aboriginal and Islander people were prepared to accept limitations on their traditional practices but would not accept responsibility for what was the underlying cause of the species loss. More importantly, the acceptance of a limitation on a traditional right was not seen as giving away rights but merely exercising those rights in a responsible manner.

This attitude was reinforced at a workshop in Kuranda in June 1994, dealing with the Queensland *Coastal Protection Bill*, a new bill proposed by Queensland to examine coastal protection. It was again signalled by indigenous groups represented at this meeting, that they were not willing to accept a white planning system which was responsible for the destruction of the coastline when traditional practices had preserved and made more valuable those areas under Aboriginal control.

5. So what of the future for Cape York Peninsula?

There is considerable resistance amongst non-indigenous people to include indigenous people in management decisions. Indeed, a continuation of the cycle of dispossession and a re-emergence of a neo-colonialist trend, is leading more and more Aboriginal people away from the path of reconciliation. There is one area in particular where the concerns about Aboriginal involvement in land management decisions is constantly being criticised. This is the mining industry.

There is considerable concern in the mining industry, and indeed as reflected in Geoffrey Ewing's paper at the symposium, that somehow Aboriginal land and Aboriginal interest in land is diametrically opposed to mining interests.

As Noel Pearson said in a speech to the Australian Petroleum Exploration Association Conference, in March 1994:
'I consistently said throughout the Mabo campaign last year that fundamental to the post-Mabo scenario was a reconciliation between the Aboriginal community and the mining industry. It is reconciliation that this country really needs as a matter of urgency, but that reconciliation will not take place if the industry is unwilling to take on board a new regime for dealing with the Aboriginal people. Because the dealing is no longer based upon charity, it's no longer based upon the presumption that all Aboriginal rights emanate from Crown largesse and Crown right. After Mabo, it is clear Aboriginal rights arise out of law. They are a matter of right, and that right emanates out of the English law that was brought on the shoulders of the colonists who landed at Botany Bay in 1788.

It is the guarantees of that law which say that Aboriginal right is not a matter of compassion or largesse, or sympathy in the polls or in the corridors of parliament. It is a matter of law, it is not a matter of referenda and how much political support there might be in the broader community for the recognition of these Aboriginal rights it arises out of a fact of law and whether the country is prepared to be subject to and obey the rule of law.'

So for the future of Cape York, we can see that Aboriginal people desiring a significant role in the management of land will also seek to influence the way that land is used. This is not confined to land but will extend to the management of sea and sea country. Aboriginal people desire strongly to be involved and to make decisions, not merely to be consulted and then for decisions to be made for them. But at the same time that does not mean that Aboriginal people wish to go it alone. There is a recognition amongst Aboriginal people of the need to include scientific and technical expertise in decision making.

The desire for Aboriginal people to be involved in decision making on land use issues is paralleled by an international awareness of the importance of indigenous people in the sustainable management of land. At a number of levels, Aboriginal people and indigenous people have been involved in developing processes and protocols for bringing traditional knowledge to support sustainable management.

In the Rio Declaration of Environment and Development (United Nations Conference on Environment and Development 1993), Principle 22 acknowledges the role of indigenous people. It states: *'Indigenous people and their communities, and other local communities, have a vital role in environmental management and development because of their knowledge and traditional practices. States should recognise and duly support their identity, culture and interests and enable their effective participation in the achievement of sustainable development.'*

Chapter 26 of Agenda 21 'A global action planned for environment and development' deals specifically with the role of indigenous people and their communities.

26.1 - Indigenous people and their communities have an historical relationship with their lands and are generally dependants of the original inhabitants of such lands…

They have developed over many generations a holistic, traditional scientific knowledge of their lands, natural resources and environments. Indigenous people and their communities shall enjoy the full measure of human rights and fundamental freedoms without hindrance or discrimination.

Their ability to participate fully in sustainable development practices on their lands has tended to be limited as a result of factors of an economic, social and historical nature. In view of the inter-relationship between the natural environment and its sustainable development and the cultural, social, economic and physical wellbeing of indigenous people, national and international efforts to implement environmentally sound and sustainable development should recognise, accommodate, promote and strengthen the role of indigenous people and their communities.'

The ideas show that the changes taking place in Cape York mirror the changes on the international scene.

6. Conclusion

The opportunity is before us to make Cape York Peninsula a model for the future. With its great environmental values and through a recognition of the vitality and importance of the Aboriginal traditional custodianship, reconciliation can receive a tremendous boost.

It is now up to non-indigenous Australians to make the necessary concessions to involve Aboriginal people.

Unless the current systems and institutions can adapt and become more accepting of Aboriginal aspirations we shall lose an important opportunity to protect one of the truly great savanna regions of the world. Whatever the future holds it must be one of co-operation and understanding. It is not for any one person to set the aims for the Cape York Region — however I hope that I have illuminated a path that we can follow.

References

Australia. Biological Diversity Advisory Committee (1993). *A National strategy for the Conservation of Australia's Biological Diversity*. Department of the Arts, Sport, Environment and Territories, Canberra.

Hawkesworth, J. (ed). (1795). *An account of a voyage around the world with a full account of the voyage of the Endeavour in the MDCCLXX along the east coast of Australia*. Edited by J. Hawkesworth from the journal of J. Cook and the papers of J. Banks, London.

United Nations Conference on Environment and Development. (1993). *Report of the United Nations Conference on Environment and Development, Rio de Janeiro, 3–14 June 1992*. United Nations, New York.

Wharton, W.J.L. (ed). (1893). *Captain Cook's journal during his first voyage round the world made in H.M. Bark "Endeavour", 1768-71*. A literal transition of the original manuscript with notes and introduction, London.

Chapter 16

Towards Responsible Land Use — Conclusions and Outcomes

Peter Huthwaite,

*Rural Management Development Centre,
Queensland University of Technology,
Box 2434, GPO Brisbane, QLD 4001
Australia*

1. INTRODUCTION

The organising committee of this symposium believed that it was important for a workshop to discuss savanna land use issues. It was envisaged that this workshop would integrate ideas from a wide representation of savanna users, policy makers and scientists on ways that tropical savanna land use and management decisions can be improved, taking into account economic efficiency, environmental integrity and social/cultural equity.

Symposium delegates were assigned to one of four workshop groups so that there was an even spread of the various land users, scientists and policy makers among the groups.

2. THE WORKSHOP PROCESS

Each group approached the task in a different way but there were marked similarities. Three groups used some sort of voting process to establish priorities. In two cases at least this voting process produced a degree of angst, disbelief in the results and then questioning of the process.

This reinforces a view I hold that the closer we get to some sort of consensus the less meaningful the outcome is because of the degree to which it has been compromised. I suggest that in this process we should avoid proposing actions and stick with general, broad direction which can be fleshed out later by sub-groups which have experience and interest in the area.

For these reasons I do not believe it is possible nor appropriate to achieve one of the originally envisaged outcomes, which was the development of a set of guidelines that could be used in policy formulation, conflict resolution or perhaps be the basis for the formation of a Working Group on tropical savanna land use.

Nevertheless I think we all agree that the processes used have legitimacy and thus the outcomes can be truly said to represent this, admittedly unrepresentative, group. This is supported by:

- the level of participation we observed throughout most of the group activities;
- the spread of representation in each group; and
- the fairness and transparency of the process in each group which has produced a truly open result.

3. THE COMMON GROUNDS

From listening to the various groups and looking at their written summaries I was able to distil the following themes which were common to all groups.

a) A commitment to sustainability in all forms of land use and land management.
b) A recognition of the legitimacy of very diverse interests, rights and perceptions about the use and management of the region. This implies the development of a consultative process which recognises this legitimacy. The consultative process needs to accommodate the difficulties caused by distance, lack of information, demands on time etc.
c) At the same time there was concern about the resolution of land tenure issues as a precondition almost for this recognition and legitimacy. The uncertainty resulting from lack of resolution could in itself negate the time and effort put into the consultative process.
d) There were broad ranging comments about education, training, appropriate information sharing, equitable access to information - both aboriginal and non aboriginal communications, issues of language and access.
e) A need for baseline data, performance indicators and ongoing monitoring to provide a starting point and then response to management and land use was considered an important area for consideration by all groups.
f) Most of the comment about science and technology bordered on the negative with comments focusing on the limitations of science in resolving many of the conflicts likely to arise. However there was a common thread of the need for science to seek out and understand the clients' needs and then to deliver them in an appropriate form.
g) This then led on to a need for codes of practice, standards and the development of an ethic in the way the region is used. Self regulation has been raised by several contributors during the symposium and this came through in the group discussions also.
h) Groups also recognised the need for a community response to national and international obligations; at the national level the moves towards ESD and at the international level agreements such as the biodiversity agreement.
i) Finally, groups generally asked the question 'is a shared vision attainable?' Often this was not expressed in so many words but each group carried some concern about the difficulty in obtaining some form of shared vision for the future of the savannas.

Can we deduce anything from this common ground? Interestingly, for a symposium dominated by scientists and technocrats there was virtually no mention of technological limitations. Most constraints were seen as difficulties in interpersonal relationships, either social or cultural, communication problems, or poor or misperceived understanding of the problems and issues.

There is clearly an opportunity in the savannas to come to terms with appropriate language, information sharing, equitable access to information and skills —

all in a region which has a long history of isolationism and difficulty in service delivery.

4. FACING THE CONFLICTS

This group has a commitment to working together — if nothing else it has cost most of us a reasonable sum of money to be here. Yet even we are in conflict about critical issues though there has been tacit agreement to cooperate.

Think, then, what a less committed group, say the wider community or a special interest group, will generate when they become involved in the group consultation process.

There is no doubt there will be conflict, there is no doubt there will be opportunities for 'win-win' solutions and there is even less doubt there will be many 'win-lose' solutions. Some people will lose and most will have real difficulty accepting that loss for the common good.

As a committed group I suggest we/you need to be prepared for the inevitability of conflict situations in which someone will lose. They will not like it, they will fight it, they will make your life miserable but at the end of the day there will be a better outcome than if we pussyfoot around and try to protect everyone.

For this reason the process, the community consultation, the development of policy has to be totally accessible and transparent. If this happens the conflict is manageable even while it continues (see Craig, and O'Donnell and Nolan, this volume).

Political reality

All groups talked about politics, about the political process and how to make it work to benefit the region. I am not a politician but have been around Government and Parliamentary Offices now for long enough to know the underlying political maxim, 'squeaky wheels get the oil.'

Northern Australia has not done badly over the years but just look at how. Who have been the real beneficiaries of taxpayer largesse. I suggest most of them have ended up back in Canberra with a promotion and lots of dinner party stories.

I believe there is a fundamental division between urban and rural populations in the same country. This is quite strong in Australia because there is so little contact between the two groups. It is thus essential to link into issues which provide these urban voters with their 'kicks'.

I do not think anyone would doubt that over the past decade two groups — aborigines and conservationists — have outperformed all other groups combined in this region in terms of influencing the political process. They are focused and coherent, appear to be united (but in reality they are no more united than the graziers) and they have a clear political agenda.

The region has less political meaning than a small handful of marginal seats in western Sydney or Melbourne. The region must identify with and harness current and future political trends if it is to get anywhere.

5. CONCLUSIONS

I do not think it is my task to draw any further conclusions from your deliberations. The range of priority issues is fairly clear, the responsibility for doing something about them rests with all of us. We will no doubt take away with us our personal agenda for action. I can tell you it has been a most useful few days of input for my landcare activities and I guess everyone will have some element of this conference, these deliberations which has significance for their current activities.

Let us not try to reach some meaningless parenthood statement, let's instead accept the diversity and get on with the job.

Chapter 17

Summative Address: Perceptions, Issues, Trends and Roles

R J Clements

CSIRO Division of Tropical Crops & Pastures,
306 Carmody Road,
St. Lucia, QLD 4067,
Australia

Abstract

*The contributions of participants in the Symposium are reviewed with particular reference to the apparent perceptions of a range of savanna users and to four key issues: production **versus** conservation; public **versus** private benefits; regulation **versus** freedom; and the existence of a wide range of values and beliefs. Perception and reality are often quite different. Also, community expectations of different land users are not consistent, and higher standards of land management are demanded from some groups than from others. However, community attitudes appear to be shifting rapidly towards a stronger conservation ethos, and significant shifts are occurring in the relative bargaining power of different groups of land users. The Australian tropical savannas do not suffer the pressures of human population that are common in other countries, and it should be possible to achieve sustainable savanna use through a mixture of private, common (i.e. group) and public ownership of land, restrictions on property rights, and regulation (including self-regulation) of land users to protect the public interest.*

1. INTRODUCTION: SOME ISSUES

This Symposium provided a learning experience for all participants. Biophysical scientists emerged with a much greater understanding of the limited role of their science in resolving conflicts about savanna use. Cattlemen, soldiers, miners, environmentalists, tourism operators and aboriginal representatives learned a great deal about other users of the land. Social scientists gained a greater appreciation of the values and perceptions of the major stakeholders in the land use debate. Critical issues were identified, and contrasting viewpoints were expounded and debated. Those who participated in the pre-conference tour in the Charters Towers district were able to inspect the effects of mining, cattle grazing, military training and tourism on land condition. Few emerged from the Symposium with their perceptions and values unchallenged. While it is too early (and perhaps over-ambitious) to claim that the Symposium was a watershed in the debate about the use of Australia's tropical savannas, six months after the event it is already clear that the participants shared an unusually revealing and positive experience.

Several underlying issues or conflicts were identified by Robyn Williams in his keynote address, and others emerged subsequently. Together they provided a constant background to the presentations and comments of the participants. They included production *versus* conservation; public *versus* private benefits; regulation *versus* freedom; and the existence of a wide range of values, perceptions and beliefs. In particular, differences in value systems are central to the debate and generate most of the heat. In my summative address, I tried to illustrate the importance of perceptions and values by pretending to be a miner, a soldier, a tourist, a grazier (cattle rancher), an aboriginal and a scientist. My intention was to illustrate how Symposium participants from a wide range of backgrounds might have reacted to the speakers. Like many theatrical productions, it relied heavily on exaggeration and caricature. My handwritten notes for this presentation still capture for me the special colour of the Symposium, and I therefore present below an edited version of my actor's lines, annotated with the benefit of hindsight and reflection. Later, I make some observations on the broader international scene in order to illustrate the complexity of the issues.

2. PERCEPTIONS, TRENDS AND ROLES

THE MINER: *Well, I think Geoff Ewing really hit the nail on the head. We generate real wealth for this country — we provide half the merchandise exports, and a lot of the jobs. We pay billions of dollars in taxes every year. And we hardly use any of the land! We spend a fortune in rehabilitating mined sites — in fact, we're the biggest employers of environmental scientists in the country — but no one gives us credit for this. Instead, they make it steadily harder for us to gain access to new sites.*

In some cases they won't even let us assess the mining potential of the land — so governments end up making decisions on land use without even understanding the options. Yet exploration has practically no environmental impact at all. They just don't want us to explore in case we find something! But even if we did, we might not tell them, because knowledge is power in the mining game.

We don't say that economic considerations always have to prevail — but we think the balance is wrong. We need to educate the community to understand the economic benefits of mining and the contribution that we are making to science.

Looking back, one of the constant refrains at the Symposium was the importance of education and access to information. Proponents of a particular land use believed that if only their opponents could be educated to understand their particular point of view, opposition would vanish. But as John Taylor and Christine Nolan pointed out, the argument usually is not about facts, but about conflicting values. The clash of core values is at the heart of conflicts over land use (Cosgrove *et al.* 1994). Furthermore, as Nolan explained, the 'experts' are rarely non-partisan; instead, they become advocates for particular causes. Trial lawyers refer to them as "saxophones" - musical instruments to be manipulated by the lawyer to produce the desired notes.

THE SOLDIER: *Community expectations of the army are greater than for anyone else — and I don't know why. We have a very high profile and a very poor image. I can't understand that. Major Joe MacDonald put our case really well. The social benefits we provide are pretty obvious. We also pump nearly a billion dollars each year into the Townsville economy, and one way or another we employ 12% of the Townsville workforce. We manage nearly a million hectares of land in northern Australia, and we really look after it well. We protect heritage sites. We allow aboriginals access to their special sites on our land. We employ an environmental officer, and we provide environmental impact statements and environment management plans. We measure indicators of sustainability — water quality, biodiversity — and our training areas are rich in wildlife. We obey all the rules, so our land is in excellent condition, and endangered species take refuge in it — and then the public wants to protect the site! They should give us more land, not less!*

One of the issues that emerged during the Symposium is that 'the rules' differ between end-users. Thus, we demand higher standards of land management from some groups (miners, military) than from others. For example, we demand that miners should rehabilitate the land they degrade. On the other hand, as Hatch *et al.* (1993) have pointed out in relation to South African graziers, many of the costs of resource degradation by Australian graziers are externalised and borne

by society. However, there are signs that this is changing. For example, the *Northern Territory Pastoral Land Act* (1992) now requires graziers not only to manage their land so as to prevent degradation, but also to participate in monitoring and improving land condition within the limits of their financial resources and available technical knowledge. The focus has shifted significantly from property development to resource conservation. While the emphasis is on voluntary and co-operative land care and management, the Act has enough teeth to enable the State to enforce improved land management if it cannot be achieved voluntarily.

There is plenty of evidence that land occupied by defence forces can provide refuge for endangered species and can enhance the conservation of ecosystems. An example at the Symposium was provided by Shaw (this volume) who described a monitoring system which has revealed the existence of a number of rare plants and animals on US Army land. In 1983, I visited the USDA's El Reno Research Station in Oklahoma, where I was shown a rare remnant of eastern tallgrass prairie that has survived on land previously occupied, from 1874, by military forces. Closer to home, participants in the tour that preceded this symposium were very favourably impressed with the condition of the pastures on the High Range Training Area and Dotswood Station, which appeared to have a more desirable botanical composition than those on neighbouring properties. Thus, the public perception or expectation that military lands are degraded may be far from reality.

THE TOURIST: *Hey, this is a really neat country. I've just been having an 'outback experience'. I want to see it all - but it's a long way between stops! I zipped in to see Katherine Gorge; but you know what? They wanted me to pay! Why should I have to pay to visit a National Park? So I burnt rubber across to Undara Lava Lodge in Queensland. I'd heard about this place — it's a really neat operation, and they have some really cute tour guides. I met a bloke called Gerry Collins, and he told me about this Savanna Symposium. It sounded different, so I came along to see what I could learn and to hear what Gerry had to say. He talked about the need for self-regulation of the savanna tourist industry. It made a lot of sense to me. There was another bloke, Peter Bridgewater, who talked about National Parks and the need for research. I've never thought about it much before, but I guess those are my research needs! And it's taxpayers' money the scientists are spending — so why can't I have more say in what gets studied? I was hoping to see some hairy nosed wombats on this trip. Now, where can I get more information about them? Why are they disappearing?*

The tourist's confusion highlights some of the most fundamental issues in savanna land use. Who benefits, and who pays? What is the balance between private and public benefit, or between present and future benefit? How can we value biodiversity? What is the balance between regulation and freedom, and how can we encourage self-regulation? How should research priorities be determined? Each of these questions deserves a proper analysis, and there was not time during this Symposium to do much more than raise the issues. However, the issues are at last being debated and researched. To take just one example, there is now a small but rapidly growing body of literature on valuing biodiversity. John Taylor mentioned briefly a number of different valuation concepts and options. These can be broadly classified as use values (production, function and option values) which can be expressed in terms of net economic returns, and non-use or preservation values (existence and bequest values) which are not easy to express in such terms. The development of methods for measuring non-use values and including them in economic analyses is a significant research challenge (Cumming 1993). If we could find a way of converting amounts or qualities on one scale of values to another, we might take a lot of heat out of the production *versus* conservation debate.

THE GRAZIER: *I was pretty impressed with what John Stewart had to say. It's a hard world, and we've had to struggle for everything we've got. We're the survivors — we've prospered in these Australian savannas for more than a hundred years. It's a never-ending battle to survive the cost-price squeeze and hang on through the droughts. New technologies have kept us in business. My grandfather wouldn't recognise the place now — new cattle breeds, sown pastures, feed supplements, new machines — and computers! And there were lots of posters at this Symposium describing new benefits for us from research. I heard Bill Winter say that scientists should be 'honest brokers' but we need them to keep on delivering benefits to us — and we're willing to pay for that! For one thing, they've got to find ways for us to control woody weeds. Land tenure is a big issue for me, and some of the things John Holmes had to say were pretty frightening. I can cope with more regulation, as long as I can stay in control of my own destiny. But environmentalists really worry me. I'm not sure if I can afford to maintain all this biodiversity. If only there was a bit of money in biodiversity, it would be a lot easier.*

Graziers in Australia's savannas have many disadvantages, but these are balanced by some significant advantages. They have a record of success, and a good deal of political and regional clout. Properties are large enough to cushion the industry against all but the most extreme climatic events and market slumps. Technological innovation has kept the cost-price squeeze manageable, and the industry is served by a highly competent and dedicated group of researchers funded mainly from the public purse. But the challenge for the beef industry, as John Taylor pointed out, is that community attitudes are changing rapidly away from a production focus, with an emphasis on economic values, towards a conservation focus, with a new emphasis on non-market values. Many environmentalists (and perhaps a significant section of the wider community) now expect farmers and graziers to maintain biodiversity on private lands, recognising that

National Parks alone are inadequate for the task. As public funds are shifted increasingly towards 'public good' research, the beef industry will have to pay more for the production research on which it will increasingly rely, and the balance between graziers and environmentalists in determining the research agenda will be altered.

In contrast to the prevailing dogma of only ten years ago (Cumming 1993) there is now plenty of evidence that pastoralists can make money from wildlife, and that domestic livestock can be combined with at least some game species. Cumming argues that the key is to grant pastoralists access rights to the natural resources on their land. In South Africa, Zimbabwe and Namibia where this has been done, the results have been striking. For South Africa, Pauw and Peel (1993) provide the following information: 10 000 game ranchers out of a total of 60 000 commercial farmers actively seek an income from their game; 3500 ranchers have erected game-proof fences; and 80% of game ranchers combine livestock and game. However, the complexity of managing a multi-species production system may be beyond most pastoralists, and the research needed to underpin management recommendations is in its infancy.

THE ABORIGINAL: *The graziers call themselves survivors after occupying this land for a few generations — but we've been here for 40 000 years! The land means everything to us. It's our life blood. The land doesn't belong to us; we belong to the land. We must re-invigorate it — but how are we to make our way? All our important gains have sprung from our own actions. White people don't understand us. They don't understand our diversity of views, and they concentrate on negative images of us. We have so much to offer, but we are plagued with negatives, and we are badly disadvantaged. We're inexperienced in modern business, we lack access to capital, and we (and the whites!) have unrealistic expectations. With leaders like Darryl Pearce, Matthew Baird and Andrew Jackson we are being heard at last. We know that knowledge is important — but what good is that knowledge to aboriginal people? Where are the answers to our problems?*

The issue of inequality in bargaining power was raised many times during the Symposium. Bill Winter, I think, raised it first during his discussion of the limitations of scientific research in resolving land use conflicts. John Taylor asserted that the group with the most power will always dominate the decision making process. Several other speakers set out the credentials of their particular interest group. However, it was Donna Craig who pointed out the importance of equality in bargaining power in resolving disputes fairly. She and Christine Nolan emphasised the importance of identifying and involving all the stakeholders. Any group that is severely disadvantaged during negotiations is likely to fare poorly and to disown the result.

THE SCIENTIST: *Well, this was a strange conference — not really what I expected. The poster presentations were great — lots of good science there — but the big issues at the Symposium didn't have a lot to do with science. I'm surprised to discover that scientists aren't 'honest brokers', but I can see now that my science isn't really value-neutral. I agree with Bill Winter about the need to develop indicators of sustainability, but I was confused by his statement that one group's positive indicators might be another group's negative indicators. And Donna Craig and Christine Nolan went pretty close to saying that we need social and cultural indicators of sustainability as well! I guess I can see the possibilities of teaming up with some of these sociologists, geographers and economists — but is it really science?*

It came as a shock to many scientists to discover that science is not central to resolving land use conflicts. Bill Winter outlined a number of situations in which the power of science to illuminate the debate is limited. Malcolm Hadley went further, and talked about the role of science and scientists. Taking into account his comments both here and elsewhere (Hadley 1993) and my own conclusions from the Symposium, scientists are being challenged to lift their game during the next decade. Industry-specific research for savanna users will still be needed, but the game has changed in a fundamental way. We need to re-assess and re-appraise our savanna resources in a context of alternative value systems. We need to turn our attention from the endless *descriptive* studies of savanna form and function, and develop the ability to *predict* the biophysical effect of management actions on savanna function, at a range of scales that are relevant to land users and policy makers. This will require an integration of process research, spatial analysis and indicators of sustainability. We also need to integrate 'hard' and 'soft' sciences — to marry together different knowledge systems — and to grapple with the complexities of diversity at the interfaces. These are challenges not only for researchers but also for research managers, because assembling and maintaining teams of such diverse specialists will require vision and determination beyond the ordinary. Lastly, whether they are 'saxophones' or not, it is possible for scientists to influence and change the attitudes of society, and they have a responsibility to do this in a balanced manner.

3. THE INTERNATIONAL SCENE

Although this was billed as an International Symposium, the prime focus was on the Australian savannas. Malcolm Hadley was the only major speaker from overseas, and eighty percent of the posters were provided by Australians. How does the international scene differ, and what can we learn from the experience of others? In answering these questions, my main sources of information apart from Hadley's paper are Young and Solbrig (1992) and the Proceedings of the XVII International Grassland Congress (1993) which contain

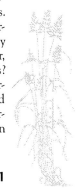

many relevant contributions, most of which I will not cite specifically.

Young and Solbrig list the following issues of significance for the management of savannas: population pressure; political instability and land alienation; misplaced government optimism; land degradation and loss of biodiversity; and lost cultural identity. The first two of these issues are less important in Australia than elsewhere. Consider, for example, population pressures that occur in other savannas. Toledo (1993) has described the extreme degradation of the South American Chaco region due to population pressure and mismanagement over a period of only 120 years. This is a region of inherently low carrying capacity, but in which 85% of the holdings are less than 2000 hectares in area, usually unfenced except for boundaries. Grazing is excessive and poorly controlled, the frequency of fires has been reduced, infestation by woody weeds is extensive, large native animals have almost completely disappeared, unsustainable cropping is widespread, and livestock production has declined. Population pressures in many parts of Africa are even more extreme, with more than 60% of the population commonly involved in agriculture or pastoralism. The contrast with Australia is striking.

People are the forgotten element in savannas. They have a major impact through their hunting, grazing, burning and cultivating activities, which are quite well understood at a mechanistic level. However, the complexity of the interactions between the specific activities of people in time and space, and the relationship of those activities to the socio-economic and cultural motivations of the people, are poorly understood.

The interaction of people with savannas is in turn intimately conditioned by land ownership. Land ownership and access issues are hardly new, and they are just as sensitive in Australia as elsewhere, as illustrated by several speakers at this Symposium. What is not clear are the conditions in which we should promote private, public or common land tenure. We need a more robust and universally applicable framework to guide our thinking. Young and Solbrig (1992) argue: *"Case studies suggest that land privatisation increases the rate of land degradation and reduces productivity"*; and *"Privatising the savanna range has not worked: neither in a social sense, nor in an environmental sense"*. Most people would find this generalisation far too extreme, but there is good evidence that individuals acting in self-interest will seek to make short-term gains by over-using their resources (or shared resources) if they do not have to bear the cost of the disbenefits to the community, particularly if future benefits are heavily discounted. Young and Solbrig recommend that responsibility for managing savannas should be devolved to local communities, who should be enabled and empowered to make a sustainable contribution to their economy.

There was some muted support for this view at the Symposium, and the Cape York Peninsula Land Use Strategy described by Matthew Baird provides a notable Australian case study. However, the security of land tenure for individuals in Australia has led to both individual and national (community) benefits. We should aspire to a system in which the *disbenefits* to the community — the small *per capita* losses by the whole community, including generations unborn — are factored into the equation. Can we achieve this through a mixture of private, common (i.e. group) and public ownership, restrictions on property rights, and regulation (by governments and groups) to protect the public interest? If we cannot, how do we proceed? In particular, how do we achieve sustainable management at successively greater scales such as catchments and regions? The starting point is to raise the awareness of land managers and policy makers to the need for collective action to protect the common interest, and the Landcare movement in Australia is already providing a mechanism for this.

My final comment relates to the affordability of conservation. There is a growing international awareness of the need for conservation of resources, and a growing willingness of countries to sign international conventions which specify commitments by participants. However, if farmers or graziers are to pay for the disbenefits inflicted on the community, or to reduce their income to protect the public interest, they can only do so if their standard of living remains acceptable. Affluent countries such as Australia may be able to afford sustainable systems, although John Holmes pointed out that returns to grazing are nearly always modest. However, poor people in poor countries often cannot afford sustainable systems, as was pointed out by several speakers at the XVII International Grassland Congress. Their first priority is survival, *today*. Even affluent Australian farmers who are unusually environment-conscious may be forced to adopt non-sustainable practices in a run of unfavourable years (Hutchings 1993). In Australia, rural reconstruction schemes and market forces may provide structural solutions to the production/conservation dilemma, but what are poor countries in the third world to do? If small-holders are displaced, how are they to survive? And if they are not displaced, for how long are they to remain at the lowest levels of economic subsistence? Can we extract payment from remote beneficiaries in the affluent world to compensate these people for reducing their use of savanna resources? These are extraordinarily difficult questions, and the answers will require us to address each of the conflicts I listed at the outset: production *versus* conservation; public *versus* private benefits; regulation *versus* freedom; and conflicting value systems.

References

Cosgrove, L., Evans, D.G. and Yencken, D.G.D. (1994). *Restoring the Land: Environmental Values, Knowledge and Action.* Melbourne University Press, Melbourne.

Cumming, D.H.M. (1993). Chairperson's summary paper. Session 57: Biodiversity in rangelands and grasslands. *Proceedings of the XVII International Grassland Congress,* Palmerston North, pp. 2103-2105.

Hadley, M. (1993). Chairperson's summary paper. Session 7: Arid and semi-arid zones. *Proceedings of the XVII International Grassland Congress,* Palmerston North, pp. 81-82.

Hatch, G.P., Tainton, N.M. and Ortman, G.F. (1993). *Costing range degradation. Bulletin of the Grassland Society of Southern Africa,* 5: 17-18.

Hutchings, T.R. (1993). Dalrye - a farm business in southern New South Wales. *Proceedings of the XVII International Grassland Congress,* Palmerston North, pp. 1957-1960.

Pauw, J.C. and Peel, M.J.S. (1993). Game production on private land in South Africa. *Proceedings of the XVII International Grassland* Congress, Palmerston North, pp. 2099-2100.

Toledo, C.S. (1993). The Chaco savanna lands of South America with particular reference to the processes of degradation in their pastoral and forestry resources. *Proceedings of the XVII International Grassland Congress,* Palmerston North, pp. 241-246.

Young, M.D. and Solbrig, O.T. (1992). *Savanna management for ecologial sustainability, economic profit and social equity.* Man and the Biosphere (MAB) Digest, UNESCO, Paris.

Index

A

aboriginal communities
    and agriculture 100
    arts and crafts 100
    attitudes to science 102
    conservation values 161
    distribution, northern Australia 11-12
    diversity among 90-92
    economic status and contribution 94-100
    and the grazing industry 95-97
        trends 152
    health and welfare 90-92
    knowledge systems, and development 140-141
    land rights 35, 92-94
        traditional 89-90
    land tenure, extent 152
    land use 109
        and alternative uses 112, 113, 171
    and the mining industry 97-98, 153
    participation in land management 162-163, 162-164
    population 90
    and protected areas 98-99
    regard for
        in development policy 122
        in regional planning 41-42
    relationships to lands 89-90
    and the tourist industry 98, 153
aboriginal settlements
    outstations 93-94
    socioeconomic characteristics 11, 14, 15
agriculture *see* cropping industries; grazing industry
arts and crafts, aboriginal enterprises 100
ASTEC *see* Australian Science, Technology and Engineering Council
attitudes
    of land user groups, mutual 110-115
    of private enterprise, to conservation 141-142
    to land uses, trends 115-116
Australian Centre for Minesite Rehabilitation 86, 152
Australian Defence Force 68-79
    see also defence ...; military ...
Australian Minerals and Energy Environment Foundation 85-86
Australian Mining Industry Council 86
    policies, aboriginal rights 98
Australian Nature Conservation Agency 2, 59-60, 157
Australian Science, Technology and Engineering Council
    report on tropical research 1-2, 23-24, 25, 58

B

beef
    markets, trends 50, 52, 151
    processing 52
beef industry *see* grazing industry
British Columbia Task Force on Environment and Economy 120
Brucellosis and Tuberculosis Eradication Campaign 32, 49

C

Cape York Peninsula 159-163
Cape York Peninsula Land Use Strategy 34, 42, 162
Cape York, Queensland
    land rent types 33
cattle
    exports, trends 50
    husbandry
        development 48-49
        research 49, 51
cattle industry *see* grazing industry
Commonwealth Scientific and Industrial Research Organisation *see* CSIRO
communications
    aboriginal enterprises 99

Index

mining industry activities 85-86
Community Development Employment Programme 94
community participation
    in conflict management 119-124, 128
    in ecosystem management 142-143, 156
    in land use planning 59
        military 76, 78
    in mining development planning 84
conflict
    between land uses 113-115, 116
    between user attitudes 110-111
    in environment issues 126-127
conflict management 116, 127, 129-130
    aboriginal issues 100-101
    in land use policy 167
    reserve management, cases 143-145
    see also mediation
Conondale Range Land Use Study 127
Cooperative Research Centre for the Sustainable Development of Tropical Savannas 2, 52, 101, 157
Council for Scientific and Industrial Research *see* CSIRO
cropping industries
    aboriginal participation 100
    development 154-155
    land use 110
        interactions with alternatives 113, 114
    see also forestry
CSIRO
    defence mobility studies 77
    land systems studies 9
CYPLUS *see* Cape York Peninsula Land Use Strategy

D

defence policy
    northern Australia, trends 69-70
Defence Regional Support Review 70
Defence Science and Technology Organisation 78
dispute settlement *see* conflict management

E

ecologically sustainable development *see* sustainable development
economic development
    aboriginal participation 94-98
    Australian Defence Force impact 76-77
    mining industry impact 82
    population factors in 12-13
    situation, northern Australia 150
    trends
        northern Australia 150-155
        world savannas 135-136
ecotourism *see* tourist industry
environmental impact assessment
    perceptions of participants 110-112
    policy, defence forces 73
    in regional planning 41
environmental disputes *see* conflict
experts, roles in conflict management 128-129
exports *see* markets

F

forestry
    aboriginal participation in 100
    limitations on 109

G

GBRMPA *see* Great Barrier Reef Marine Park Authority
government policies 1-2
    defence deployment 70-72
    development, mediation in 129
    grazing industry, issues 51
    and land tenure 30-31
    Northern Territory 36-37
    in regional planning 41
    and research 25
    resource assessment process 123
    tourist industry 64-65, 67
    see also institutional frameworks
grazing industry
    aboriginal participation 152
    economic development, trends 151-152
    government policy issues 51, 53
    historical development 48-49
    interactions with alternative enterprises 112, 113, 170-171
    land tenure, trends 29, 35-37, 48-49, 51, 151-152
    land use 109
    legislation affecting, Northern Territory 155
    property sizes 49
    relationships with aboriginal communities 95-97
    as resource use, evaluation 7, 23-24
    sustainable development issues 50-51
    see also beef; livestock
Great Barrier Reef Marine Park Authority
    aboriginal participation 162
    model of regulatory body 155-156
    model for research policy 25
Gulf Local Authorities Development Association
    tourism policies 65-66
Gulf Region Land Use and Development Study 34

H

horticulture *see* cropping industries

I

Inbound Tourism Organisation of Australia 66, 67
indigenous people *see* aboriginal communities
institutional frameworks
    in ecosystem management 142-143
    nature conservation 59-60
    reform, needs and trends 34-35, 142-143
    resource management 155-156
    tourist industry 65-66

K

Kimberley Region Plan Study Report 34

L

land degradation
    definition and measures of 22-23
    issues, grazing industry 50-51
land rehabilitation
    areas, maintenance 56-57
land rent
    distribution, Cape York, Queensland 33
    grazing industry 51
land rights *see* property rights
land tenure
    grazing industry, trends 29, 35-37, 48-49
    marginal lands, reform 37-39
    reform, needs and trends 35-40
    traditional aboriginal views 89-90
    types, distribution, northern Australia 9, 30
    see also land rent; pastoral leases; public ownership
land types
    resource values 32-34
land use
    alternatives 108-110
        evaluation 7-10
    research in 23-24
    user perceptions 110-112
        interactions between 111-115
    conflicts 113-115
    conservation issues 55, 56-57, 72-76
    multipurpose, policy trends 36-37, 39-40
    patterns, defence forces 72
    traditional aboriginal 92-93
land use planning
    aboriginal participation 162-164
    Cape York Peninsula 162-164
    policy issues, consensus summary 166
    in regional development 41
land use surveys, availability and utility 9-10
Land and Water Resources Research and Development Corporation 2, 157
Landcare movement *see* National Landcare Program
livestock
    population distribution, northern Australia 9

M

marginal lands
    resource values 32
    tenure, reform 37-40
markets
    beef industry 50, 52, 151
    tourist industry 63-64, 110
Meat Research Corporation 50
mediation
    definition 127
    processes 121-122
    in public policy development 129
    role of experts in 128-129
    in sustainable development policy 122-124
        case studies 127-128
military activities 68-79
    and aboriginal communities 99-100
    as land use 109, 154
    interactions with alternatives 113, 114, 169-170
military bases
    distribution, northern Australia 69-70
    functional deployment 70-72
military personnel
    deployment, northern Australia 70-72
    social impact 78
military training
    land use and protection 72-76
mineral exploration
    restrictions applied to 82-83
mining industry
    economic development, trends 152-153
    land use 110
    processing component, development 153
    public communication channels 85-86
    relationships with aboriginal communities 97-98, 113, 169
    relationships with other land uses 113, 115
    restrictions on, protected areas 84-85
    socioeconomic impact 82
    and sustainable development 81-82, 152
mining settlements
    socioeconomic characteristics 14, 15-16
Murray Darling Basin Commission
    model of regulatory body 155-156

N

National Landcare Program 50
    model of development management 156
National parks *see* protected areas
National Strategy for Rangeland Development 2, 51, 52, 86
nature reserves *see* protected areas
North Australia Beef Research Council 51

O

Office of Northern Development
    research policies 157
opinions *see* attitudes

P

pastoral leases
    evaluation 32-34
    land tenure, trends 35-37
    multipurpose policies 36-37
pastoralism *see* grazing industry
perceptions *see* attitudes
political decisions *see* government policies
population
    aboriginal communities 90
    composition and distribution, northern Australia 11-12
    growth centres *see* urban development
    military component 70-72, 78
    socioeconomic characteristics 12-13
    see also settlements
Prime Minister's Science and Engineering Council
    working group on savanna research 2
private enterprise
    and environmental issues 141–142

Index

property rights 35-40
   aboriginal, and mining activities 97-98
   traditional aboriginal views 89-90
protected areas
   aboriginal participation 98-99
   aboriginal reserves 91
   conflict management in 143-145
   definition and selection 55-56, 84
   land use 109
      interaction with alternative uses 112, 114
   management, research 58-59
   and mineral exploration 82-85
   sustainable development 57-58
public ownership
   and multiple land use 39-40
   trend to 29-30
public participation *see* community participation

R

rangelands
   development 2, 51, 52, 86, 127, 145
   resource values, trends 31
regional development
   and resource re-evaluation 31-34
regional planning 34
   implications for aborigines 41-42
   needs and policies 40-42
research
   aboriginal attitudes to 102, 140-141
   in cattle husbandry 49, 51
   in defence support 78
   federal government policies 1-2
   for the grazing industry 51, 52-53
   management 24-25, 157
   policies, savanna development 136-141, 157
   for protected area management 58-59
   and resource management 20-27, 171
Resource Assessment Commission 122-123
resource development
   international aspects 171-172
   legislative regulation 155-156
   management, participatory 156
   sociopolitical context, world trends 134-146
resource evaluation
   criteria, rangelands 31
   modern trends 6-10, 29-34

S

Savanna Guide service 65-66
science *see* research
settlements
   characteristics and distribution, northern Australia 13-17
   defence personnel influence 70-72, 78
   see also aboriginal settlements; mining settlements; urban development

social participation *see* community participation
sustainable development
   definition and principles 120-121
   federal government policies 1-2
   grazing industry, issues 50-51
   indicators of sustainability 22-23, 101-102
   landscape maintenance 57-58
   mining industry aspects 81-82
   policies, Australian Defence Force 72-76
   and protected areas 59
   tourist industry, issues 64

T

tourist facilities
   development, case study 66-67
tourist industry
   development trends 63-64
   economic development, trends 153-154
   as land use 109-110
   operatives, training 65-66
   relationships with aboriginal communities 98, 113
   relationships with other land uses 113, 170
   significance of Outback concept 63-64
training centres
   environment management, armed forces 73-74
training programmes
   military 72
   tourism operatives 65-66
transport
   aboriginal participation 99
   planning, defence forces 77-78

U

Undara Lava Tube
   tourist attraction 66-67
urban development
   and community attitudes 110-111
   defence forces role 70-71
   regional resource values 32, 34
   socioeconomic significance 17

V

vegetation types
   distribution, northern Australia 8

W

water use
 conservation issues 55
weeds
   impact, grazing industry 50
World Heritage areas *see* protected areas